Innovations in Safety Management

Innovations in Safety Management

Addressing Career Knowledge Needs

Fred A. Manuele, CSP, PE
Hazards, Limited
Arlington Heights, IL

WILEY-INTERSCIENCE
A JOHN WILEY & SONS, INC., PUBLICATION
New York · Chichester · Weinheim · Brisbane · Singapore · Toronto

This book is printed on acid-free paper. ∞

Published simultaneously in Canada.

For ordering and customer service, call 1-800-CALL-WILEY.

Library of Congress Cataloging-in-Publication Data is available.

0-471-43959-2

Printed in the United States of America.

10 9 8 7 6 5 4 3 2 1

To My Family

Contents

Preface

The need for a book fulfilling career knowledge needs that has been addressed minimally by others arose from a survey made by this author of corporate and divisional safety directors. They were presented with the following scenario to obtain their perceptions of current and future knowledge needs:

> You are planning a gathering of safety practitioners for educational purposes, and you want to have workshops on those topics for which you believe additional knowledge is required, considering current needs as well as looking ahead three to five years. What topics would you choose?

The results of this survey were both fascinating and surprising. When the data were summarized and a list produced of the seminars and workshops that interviewees thought were necessary, first priority was given to the need for a better understanding of the fundamentals of financial management. Safety professionals thus wanted to better understand the meanings and significance of the terms commonly used in business meetings, and executive decision making. How to establish one's value and communicate statistics in a language that management understands received second priority.

The first two chapters cover the basics of financial management and the communication of incident statistics in financial terms.

Chapter 3, on the theoretical ideal for optimum safety performance, was written in response to the request for further information on successful safety ventures.

Chapter 4 addresses a hierarchy of controls within tried and proven problem solving techniques to become The Safety Decision Hierarchy. This is entirely new work.

The safety practitioners interviewed knew that the author had participated in the creation of an Institute for Safety Through Design, and they presented a challenge of sorts. If knowledge about safety through design was important to the careers of safety practitioners, what subjects would be included in educational seminars? Thus, Chapters 5 and 6 were written to establish why safety practitioners need to know about hazard analysis and risk assessment, and to provide a primer on these subjects.

Chapter 7, on how to avoid bringing hazards into the workplace, is a logical extension of Chapters 5 and 6. It provides practical guidance to safety professionals.

Chapter 8 presents a method to extend a basic safety function—task analysis—to also address productivity, cost efficiency, and quality, and to further establish value. The origin of this chapter is the immensity of the data on ergonomics applications that have also resulted in productivity and cost improvement.

Since the use of risk-scoring systems is becoming prevalent, the subject was comprehensively researched and written about in Chapter 9. A new three-dimensional method is presented as an answer to emerging needs.

All the safety practitioners interviewed had an interest in behavioral safety and were interested in a seminar on current developments in that field. Chapter 10 speaks to that interest.

Chapter 11 provides a history of the safety through design movement, about which safety practitioners should have a fundamental knowledge.

Beyond a doubt, the practice of safety is in transition. This book addresses some of the new knowledge needs.

FRED A. MANUELE, CSP, PE

Acknowledgments

Critiques of the chapters in this book by colleagues were greatly beneficial, and these individuals deserve recognition. I express my thanks especially to Wayne Christensen, and to Thomas Cecich, Michael Douglas, Jim Howe, Edward Kornas, Gary Lovested, Bruce Main, Kristen McMurphy, Timothy McNamara, William Montante, John Polhemus, Thomas Selders, and Harry Taback.

Introduction

An abstract is provided for each chapter to serve as a ready reference on content for all to whom the book will be of value: safety practitioners, educators, and students.

1. A Short Course on Financial Management

Chapter 1 was written in response to the expressed need of safety practitioners for an understanding of the basics of business financial management, the language of finance, the budgeting process, the impact of adequate or inadequate cash flow, performance measures, the primary elements in a financial report, and how executive decisions are made. With this knowledge, safety practitioners feel they will be more effective in assisting their companies attain established goals.

2. Establishing Your Value: Communicating Incident Statistics in Financial Terms

Safety practitioners must realize that management relies on the statistical outcome of safety-related initiatives as a measure of performance, and that finance is the language of management.

Chapter 2 has three parts. Part One gives examples of financially related statistical analyses and presentations made by safety practitioners. Part Two is devoted to computing the financial implications of direct and indirect hazards-related incident costs. Part Three discusses other aspects of performance measures and their financial relationships.

3. The Theoretical Ideal for Optimum Safety Performance

Chapter 3 responds to the following question often posed by safety practitioners: How can we do it better and, at the same time, present a framework that should be considered by those who draft a safety management standard. A model is presented that emphasizes the importance of culture, which then creates a system of expected behavior that, in turn, translates into superior policies, standards, and procedures. This results in an inherently safer workplace.

4. The Safety Decision Hierarchy

Chapter 4 represents entirely new work. The literature contains fragments of information on what is called the hierarchy of controls. This chapter examines the decision making for risk elimination and control within the context of the principles of good problem-solving techniques. It is, in a sense, a dissertation on an orderly thought process for the elimination of safety problems. A model for the safety decision hierarchy is presented, and its origins and justification are discussed.

5. Why You Need to Know About Hazard Analysis and Risk Assessment

Safety standards and guidelines issued in recent years or in the process of development require or propose hazards analyses and risk assessments. This is a significant evolution. It has an important bearing on the knowledge needs of safety practitioners. Relevant standards and guidelines are reviewed in Chapter 5. Taken in their entirety, they are convincing predictors of the future and underscore the need for safety practitioners to know the fundamentals of hazard analysis and risk assessment.

6. A Primer on Hazard Analysis and Risk Assessment

Chapter 6 provides basic information on hazard analysis and risk assessment that will be sufficient, it is hoped, to serve the knowledge needs of safety practitioners. The following topics are explored: what a hazard analysis is, how a hazard analysis is extended into a risk assessment, the steps to be followed in conducting a hazard analysis and risk assessment, descriptions of the most commonly used techniques, and examples of two-dimensional risk decision matrices.

7. How to Avoid Bringing Hazards Into the Workplace

In Chapter 7, we discuss just how involved safety practitioners are in the design and equipment acquisition processes that serve to avoid bringing hazards into the workplace. The level of involvement varies from none to the near-theoretical ideal. A composite is provided of the procedures in place at several companies to achieve hazard avoidance and control, and the procedures to be followed before new or modified equipment is released for operation. Also, a general checklist is included to serve as a reference from which a specifically tailored checklist can be developed for use in initial design reviews and for equipment acceptance. An appendix sets forth the near-ideal equipment design philosophy and practice in place at one major corporation.

8. Task Analysis: For Productivity, Cost Efficiency, Safety, and Quality

The idea for Chapter 8 arose during a review of data describing how successful ergonomic applications also impacted favorably on productivity, cost efficiency, and sometimes quality. A basic safety function, task analysis, is extended to include productivity and cost efficiency, or quality, and thereby also provides safety practitioners with an opportunity to further establish their value.

9. Risk-Scoring Systems

For many years, the typical practice has been to establish risk levels based on determinations of two factors: probability or event occurrence, and the severity of harm or damage that could result. Translating risk assessments into numerical risk scores is not necessary with such an approach. However, the world of risk assessment is in transition. Systems now in common use may be three or four-dimensional and require numerical risk scorings.

Chapter 9 reviews the origins of current transitions; the cautions to be considered in using three- and four-dimensional systems; three-dimensional, numerical risk-scoring systems in use; extended three-dimensional risk-scoring systems, which include a method to justify the risk amelioration costs in relation to the amount of risk reduction to be attained; and a four-dimensional risk-scoring system, now experiencing limited use, but for which broader use has been proposed.

A newly developed three-dimensional system is presented that can be applied for all risks of injury or damage that a safety practitioner is likely to encounter.

10. Current Developments in Behavioral Safety

In 1998, the American Society of Safety Engineers held a seminar on behavioral safety, and the presentations of several participants in that conference serve as the benchmark for Chapter 10. Since 1998, the content of the writings and speeches of these same speakers has been markedly different.

We review current acceptance of the following by some of the principle players: Behavioral safety is but one element in an overall safety initiative; for most hazard-related incidents there are multiple causal factors; application of the hierarchy of controls is the most effective means of reducing risk; and W. Edwards Deming got it right when he wrote that "performance does not come from the individual ... but from the system." These developments are remarkable when examined in the context of earlier widespread beliefs in the field of behavioral safety.

11. The Safety-Through-Design Movement

Chapter 11 takes a global view of an emerging safety through design movement. The international harmonization of safety and quality standards is discussed, driven as it continues to be by European influences.

The following are presented: the adoption of safety through design principles by American companies; the benefits to be obtained through the application of safety through design principles; the actual safety through design model; emphasis on how applying safety through design methods avoids high cost or retrofitting; and many examples of how the principles may be applied in the real-world practice of safety.

Appendix

Reference is made in Chapter 9 on risk scoring systems to the early work of William T. Fine. His seminal paper entitled, "Mathematical Evaluations for Controlling Hazards," is a valuable resource. It is, however, not easily obtained, and so for this reason, the appendix here contains a slightly condensed version of it.

1

A Short Course on Financial Management

INTRODUCTION

Why does this book contain a chapter on the basics of financial management, and what is its purpose? The idea for it originated in a study initiated by the author to identify emerging knowledge needs in the practice of safety.

A part of the study included interviews with senior-level safety professionals to obtain their comments on the subjects they considered significant in maintaining professional performance and for career enhancement. The following scenario was presented to 17 divisional and corporate safety practitioners, and 2 university professors:

> You are planning a gathering of safety practitioners for educational purposes, and you want to have workshops on those topics for which you believe additional knowledge is required, considering current needs as well as looking ahead three to five years. What topics would you choose?

The results of the ensuing survey were both fascinating and surprising. When the data were summarized and a list was produced of the seminars and workshops that interviewees thought were necessary, the need for a better understanding of the basics of financial management was given first priority.

The safety directors interviewed had management positions. They had become aware that they could be more effective in assisting their companies attain established goals if they had a better understanding of the basics of financial management. They said that they would be more comfortable in their jobs and in business meetings if they had knowledge of basics such as:

- The language of finance and financial management principles
- How corporate budgets are prepared and managed
- The impact of adequate or inadequate cash flows
- The performance measures applied to managers
- The primary elements in a financial report
- How executives make decisions when resources are limited and they have to evaluate multiple expenditure requests

With this knowledge, safety directors say they can more effectively draft their hazard prevention and control proposals, emphasizing their favorable financial impact on overall operations, including applicable aspects of productivity and cost efficiency, as well as risk reduction.

Comments made by safety directors prompted the research for this chapter. Its purpose is to address their expressed needs.

AN EMERGING AWARENESS

As the following examples indicate, other management personnel feel that they would better perform their jobs if they had general knowledge of widespread business practices and the principles of financial management. A publication that addresses the subject is the "1998 White Paper, Loss Control 2000" issued by the American Society of Safety Engineers (ASSE). The White Paper was produced as the result of a panel discussion hosted by the Risk Management/ Insurance Division to discuss the future of the loss control profession. It gives the results of those discussions, some of which relate to changes in the business environment that impact on loss control services. The following are a few of the subjects discussed in advising loss control personnel:

- Learn more about business management and operations
- Understand finance and accounting principles
- Cost/benefit and your worth to your employer

Further, consider this scenario: A president of a major division in a large and well-known company sensed a need to increase the business acumen of the management staff. Through a teleconferencing session, to which the senior safety personnel were invited, every line of the company's financial report was reviewed to assure that entries were understood and how the operations of each internal entity impacted on results.

Michael Muckian, in *The Complete Idiot's Guide to Finance and Accounting*, supports the idea that the operations of each internal entity impact on results. Consider these excerpts:

- It doesn't matter what your professional responsibilities are, your management responsibilities contribute directly to your company's profitability.
- Understanding how the company's finances work is simply good management practice.
- You'll be better able to manage your department if you understand the financial implications of your actions.
- A basic knowledge of accounting practices and principles is worth the effort that it takes to learn them (4).

BEING PARTICIPANTS IN ACHIEVING MANAGEMENT GOALS

In several interviews, concerns were expressed about the significant changes that have occurred in the business community, the greater expectations management has of safety practitioners, and their need to be perceived as providing value.

In *On The Practice Of Safety* (p. 1), the author set forth in the first chapter, "Transitions Affecting the Practice of Safety", what he perceived to be the paramount requirements of those having responsibility for operations, those with whom safety practitioners must interact in meeting their goals. Reference was made to two highly respected writers, Peter F. Drucker and W. Edwards Deming. Both of these individuals have had a major influence on the author's understanding of management concepts and practices.

Drucker, in *Post-Capitalist Society*, wrote this:

> Economic performance is the first responsibility of a business. (Without economic performance) a business cannot discharge any other responsibilities, cannot be a good employer, a good citizen, a good neighbor (101).

Drucker makes essentially the same point as W. Edwards Deming did in his often cited "Condensation of the 14 Points of Management" in *Out of the Crisis*. Deming is world renowned with respect to quality management. He proposed that an entity should establish:

> Constancy of purpose toward improvement of product and service, with the aim to become competitive and to stay in business, and to provide jobs (23).

As safety practitioners, we must understand and function within the directives given to management to achieve economic goals, to be competitive, to stay in business, and to provide jobs. Demands imposed by a highly competitive economy for more effective management performance are not going to diminish. They compel safety practitioners to be perceived as problem solvers whose counsel is sought because of their successes as participants in achieving management goals.

GOALS TO BE ACHIEVED BY SAFETY PRACTITIONERS

Without question, the more demanding business climate now existing requires greater management effectiveness, a part of which is a stringent measurement of the contributions of individuals toward achieving entity goals. For some, what is proposed here as goals to be achieved by safety practitioners are substantially different from commonplace perceptions of their roles. Safety practitioners should aspire to be perceived as:

- Participants in effectively and economically applying available and limited resources to eliminate or reduce risk so that the greatest possible good to employees, to employers, and to society is obtained from those endeavors
- Contributing effectively to productivity, cost efficiency, and quality management, in addition to safety
- Active participants in achieving management goals

WHAT'S HAPPENING: A PREDICTION FO THE FUTURE

Excerpts taken from Johnson & Johnson's *Environmental, Health and Safety Report 1998* follow. They are examples of accountability that are the norm in a small number of businesses, but can be expected to become more prevalent in the near future.

The first excerpt indicates how Johnson & Johnson has established accountability for costs and benefits through a financial accounting model. The second excerpt establishes that the strategic objectives for safety and industrial hygiene include influencing all subsidiary companies to integrate safety into their business plans, and maximizing business results for the operating companies.

Financial Accounting Model

In 1998, cost accounting models were developed to capture environmental and safety costs associated with our products and operations. In addition, the models provide companies with useful business integration tools to communicate the cost, value and return on investment of our environmental and safety activities. These models, which are being deployed in 1999 to our worldwide operations, will help identify new opportunities to improve environmental and safety performance (8)....

Safety & Industrial Hygiene—Strategic Objectives

Safety and industrial hygiene are corporate values embodied in Our Credo. They are transformed into workplace reality today at every Johnson & Johnson company throughout the world.

Johnson & Johnson sets five strategic objectives in partnering with our global businesses to achieve an injury-free workplace:

- Drive safety and industrial hygiene through design.
- Implement preventive processes that protect our employees and businesses.
- Influence all companies to integrate safety into their business plans.
- Anticipate and influence potential regulatory issues.
- Maximize business results for the operating companies (22).

What is the message here? Safety practitioners should be prepared to assist in determining "the cost, value and return on investment of . . . environmental and safety activities" and in "maximiz(ing) business results for the operating companies."

In many companies, there is a growing expectation that management personnel at all levels have an understanding of the financial measures currently applicable in their company and of the company's financial status at given points in time. With this knowledge, it is expected that managers will be able to contribute more effectively to achieve the company's goals.

Finance is the universal language of business—of its goals, objectives, and results. Variations of that theme appear in all the literature reviewed by the author. We must thus recognize that performance is measured in financial terms.

UNDERSTANDING FINANCIAL DECISION MAKING

For many of the decisions made by management that result in their spending or not spending money, one or more techniques of financial analysis may be used. An excellent source that may convince safety practitioners that it is advantageous to understand financial decision-making techniques is *Safety And the Bottom Line* by Bird and Davies. In the third chapter entitled, "Safety Applications Of Economic Principles" (79), Bird and Davies make the case that, over time, safety practitioners may succeed in securing executive support of safety activities if they persistently apply certain economic principles (axioms).

The axioms pertain to management responsibilities and the economic considerations managers apply to situations brought to their attention. The substance of these axioms, which follow, will help safety practitioners understand the importance of learning the basics of the mathematical systems used in financial decision making:

1. *The Management Control Axiom.* The control of accidents is the best way to control their cost and management has the bulk of this control.
2. *The Axiom of Economic Association.* A manager will usually pay more attention to information when expressed or associated with cost terminology.
3. *The Axiom of the Critical Few.* The majority (80%) of any group of effects is produced by a relatively small (20%) number of causes.

4. *The Key Advocate Axiom.* It is easier to persuade decision makers on matters of cost when at least one person within their own circle believes in the proposal well enough to champion the cause.
5. *The Past Forecasts the Future Axiom.* Unless positive steps have been taken to correct or control past loss experience and if related circumstances continue to exist, you should expect to have similar, if not more costly, experiences in the future.
6. *The Mutual Interest Axiom.* Projects and ideas involving costs are best sold when they bridge the wants and desires of both parties.
7. *The Economic Priority Axiom.* A manager will usually give priority to items with the potential for the greatest proportion of results from the least investment of available resources.
8. *The Vested Interest Axiom.* Managers are predominantly interested in those economic considerations affecting their own budgets.
9. *The Axiom of Adequate Information.* The timeliness of managers' decision making is directly related to the adequacy of information they have upon which to act.
10. *The Axiom of Dimensional Value.* The degree of management attention is directly related to the size of the problem.

What do all these axioms suggest? The probability of favorable action on a proposal is greater if is accompanied by cost-benefit data, if it is supported by a champion, if it addresses management's financial and budget needs, if alternatives and their costs are presented, and if it is timely.

A BRIEF REVIEW OF THE LITERATURE AND AVAILABLE COURSES

The author's work experience includes having managed a 225 person consultancy and being held accountable for achieving the performance set forth in the annual business plan. Interpreting monthly financial reports involved acquiring an understanding of the significance of budgeting, revenues, cash flow, expenses, margins, and net operating income. That experience, in turn, led the author to a clear understanding of some of the basics of financial management. It seems that in the prevailing business climate, the financial knowledge needs of safety practitioners are even broader.

Fortunately, there are several courses of action available for safety practitioners who choose to expand their knowledge of the basics of financial management. Publications on the subject are numerous. A review of some follows: They vary greatly as to depth and purpose.

Two college-level texts were reviewed for background in preparing this paper. One is *Fundamentals of Financial Management* by Van Horne and Wachowicz; the other is *Essentials of Managerial Finance* by Weston, Besly, and Brigham. As expected for college courses designed for those who must develop extensive knowledge of financial principles, they cover all subjects in ultimate detail. They are listed in the end-of-chapter references, should a safety practitioner desire additional information about financial management.

Also, the existence of such an extensive body of literature gives rise to the thought that some safety practitioners may want to enroll in what is frequently a three-hour, one-semester college course in financial management. Safety practitioners employed in nonprofit entities could consider *Financial Management For Nonprofits* by Shim and Siegel (also listed in the references) for educational purposes.

Other texts, written essentially for nonfinancial persons, are available. Four such books are:

- *What Every Manager Should Know About Financial Analysis* by Alan S. Donnahoe
- *Understanding Return On Investment* by Friedlob and Plewa
- *Finance For The Non-Financial Manager* by Herbert T. Spiro
- *The Complete Idiot's Guide to Finance and Accounting* by Michael Muckian

There are many short courses and self-study courses on the basics of financial management available: Only two are cited here. The American Management Association (AMA) offers a three-day course entitled, "Fundamentals of Finance and Accounting for Nonfinancial Executives." It was offered 69 times the year 2000. This surely indicates that many nonfinancial management personnel have a need to understand financial principles. A Self-Study course, also titled "Fundamentals of Finance and Accounting for Nonfinancial Managers," is available through the American Management Association.

A PRIMER

In this chapter, only brief comments are made on a subject about which many, lengthy works are written. What follows is merely an introductory-level response to the knowledge needs mentioned in interviews with safety professionals. Generally, the books cited here cover the same subjects, mostly but not exactly, the differences being in the amount of detail and individual author views.

THE FINANCIAL MANAGEMENT FUNCTION

A brief but good description of the financial management function appears in Spiro:

> Finance engages in two primary types of functions for top management: recording, monitoring, and controlling of financial consequences of past and current operations; and acquiring funds to meet current and future needs (1).

If a company is large enough and employs a controller, the recording and monitoring functions would be his or her responsibility. Acquiring funds to meet current and future needs and interfacing with financial institutions would be the responsibility of the treasurer, if the company has one. In any case, the chief financial officer of a company, the person responsible for the overall finance and accounting function, whatever the title used, is a person of importance and power.

Borrowing from Spiro again:

> Managerial accounting is the term used to describe the control aspects of the finance function. It deals with the preparation of financial forecasts, the development and monitoring of performance budgets, and product costing. These endeavors require considerable reliance on the historical data and expenditure patterns developed by financial accounting. In discharging these responsibilities, financial management forecasts utilize the framework and tools of economics and financial analysis (2).

ORIGINS

Several authors indicate that the concepts on which financial management is based derive from economic theory. The financial measurement systems commonly used are those applied in economic analysis for economic activity, whether for profit or nonprofit entities. For either type of operation, Generally Accepted Accounting Principles have evolved for the recording of financial transactions and for the allocation of resources. These principles are set forth in guidelines issued by the Financial Accounting Standards Board.

Over time, regardless of an entity's purpose, survival requires that revenues exceed expenses. For the entity that is to make a profit, economic investment theory prevails.

INVESTOR EXPECTATIONS: RISK TAKING

Investors, logically, expect a rate of return commensurate with the risk assumed when they provide their money to a company that operates in a competitive environment, and which may or may not make a profit. The extent of that business risk is determined by evaluating the uncertainties of the company achieving the successes that it expects to attain.

To influence investors favorably, the entity seeking outside investment must produce a convincing business plan that specifies the actions to be taken to achieve the expected profits. The business plan would include the necessary display of financial data.

Understandably, investors would be reluctant to invest in a company if its profit expectations were no greater than that of a virtually risk-free investment in federal government bonds, or an investment in tax-free, insured municipal bonds. Investors will expect a return commensurate with the risks of the business proposal under consideration. At some time in their history, almost all companies will borrow money from investors, who will closely review their business plans and budgets.

BUSINESS PLANS

What is a *business plan* and what should safety practitioners know about them? This description may be found in Muckian:

> The business plan is the where, when, how, and why of what a company must do to implement its strategy. The business plan offers general and specific guidance on reaching company goals. It outlines actions. It also separates the dreamers from the doers. An annual plan is operational and necessary to manage the company's economic needs for the coming year. A five-year plan is more strategic and designed to chart the firm's direction (16).

Operations of every entity within a company affect the firm's overall business plan and, in turn, are affected by it.

PERFORMANCE BUDGETS

A *budget*, says Spiro, is a quantitative expression of a contemplated plan of action:

> Both explicitly and implicitly, the budget, presents the intentions and objectives of management to all echelons of the organization and provides a vehicle for monitoring the implementation of plans. In addition, it enables management to assess the adherence of individuals and organizational components to the goals stipulated in the plan and thereby to provide a quantitative basis for measuring and regarding individual and departmental performance (109).

Because of the need to collect data from each of the operating entities within a company, the budgeting process is usually started early in the third quarter of a year and it often takes quite some time to finalize. Budgets will typically include:

- Sales or revenue forecasts for the coming year
- Costs for production or operations, and other expenses
- Expected capital expenditures
- Financing needed and how it is to be obtained
- Cash flow projections
- A projected balance sheet, an income statement, and a concluding indication of profit or loss

Most safety practitioners become involved in budgeting at some time in their careers. They will be more successful in such endeavors if they are knowledgeable about the company's business plan, and the basics of financial management.

PROFIT CENTERS

Why are *profit centers* established? Friedlob and Plewa give a good response to that question:

> Company managers must show they can manage the owners and creditors investment and produce the profits owners and creditors demand. Because top management must meet the profit expectations of company owners, profitability goals are passed down to lower levels. The end result is that (profit center) managers are charged with profitability goals (3).

A profit center is an organizational unit responsible for the generation of revenue and the control of costs for which financial statements can be prepared. What are the performance measures placed on managers: performance measures that determine their salaries, bonuses, and promotion or dismissal potential? Profit center managers are held accountable for maximizing profits in their sphere of influence. Profit centers are established to provide managers an operational freedom similar to that available to the entrepreneur, within the limits of the manager's and the overall business plan and budget. Nevertheless, profit center managers usually have a great deal of freedom in allocating resources and the expenditure of money.

COST CENTERS

Most safety units operate as *cost centers*, which are described as organizational units responsible for the control of costs but not the generation of revenues. A few safety units are required to sell their services within their companies and operate as profit centers. Their staffing is dependent on the purchases of their services from internal profit centers.

FIXED CHARGES

Expenses fall into two major categories: fixed and variable. *Fixed expenses* are those that remain the same, irrespective of production or sales volume. They are called overhead and represent a special concern to management. These are the expenses that must be covered, in good times and bad, and they determine the break-even point of the company. Thus, all good managers are expected to work hard to control, and if possible, reduce, all the factors that enter into overhead.

Fixed costs include expenditures such as the rent, debt payment obligations, staff salaries, and perhaps property insurance premiums. Variable costs, such as the costs of materials, costs of goods sold, and labor costs, will increase or decrease in accord with sales or business volume.

Overhead is a burdensome reality. The cost of a safety staff is generally included in overhead, except in the few cases where the safety staff operates as a profit center. Thus, it (among other vital services) is susceptible to cuts or even dissolution when management deems it necessary to reduce overhead in the interests of improving financial performance.

If the safety staff has been able to forcefully show that it provides value, it will be less susceptible to elimination. Safety practitioners therefore have to care about their company's financial status. As was said by a highly successful executive, whose education and early employment were in finance but who is now chief operating officer of an operating division:

> I wish the safety staff could walk in my moccasins for a few days so that they could understand how financial implications impact on my decisions. And they could be more convincing in establishing that they provide value to this operation. I think they do, but they don't seem to sense the importance of making their case.

CASH FLOW

Donnahoe succintly covers cash and cash budgets:

> Corporations must have cash to meet their various needs, and they must get it either from cash generated by operations or by borrowing or by additional equity from investors. So cash flow is a vital subject for corporate managers. Cash is tied up in accounts receivable ... and inventory, and well-managed companies keep them at a minimum in relation to sales volume (96).

Cash flow is thus the flow of money in and out of a business. That is all. It is not the profit made on sales. If cash flow difficulties arise, obviously a company will have to change the way it is doing business. If cash flow becomes a problem, two principle courses of action are typically considered: an increase in sales, or a reduction in expenses.

In addition to the overall budget, many companies have a cash budget as well. It is the accounting document delineating expected cash flow.

How important is the control of cash flow? Spiro gives cash flow an exceptionally high status in financial management:

> Of the many feasible supporting budgeting schedules, none is as important to the financial manager as the cash budget. The reason is simple: regardless of the future potential of the organization, unless cash is available to meet near-term obligations such as payroll, power, telephone, notes due, and so forth, the organization

> is effectively out of business. While in the private sector the organization's long-term survival depends on profitability and in the public sector on outside funding, near-term survival in either case necessitates the availability of necessary cash to meet immediate requirements (120).

How can the safety staff be affected by shortages in cash flow? In many businesses, cash flow is uneven from month to month or quarter to quarter. Shortages can affect management decision making concerning monies that can be spent, on what, and when. Recommended safety expenditures will be among the several proposals requiring cash that a manager has to consider at a given time. Being a part of the management team and having knowledge of short-term cash flow difficulties helps the safety staff better understand the final decision making and determine the most advantageous time to make its own proposals.

BALANCE SHEETS AND INCOME STATEMENTS

Authors of all the texts cited here devote considerable space to *balance sheets* and *income statements* because they are so important in determining the financial well being of a company. Spiro writes:

> The Balance Sheet and the Income Statement are key documents of financial accounting—the activity concerned with the recording and portrayal of historical financial events (32).

An example of a balance sheet, Exhibit 1, and an income statement, Exhibit 2, are provided. They are here to give the reader additional "primer" knowledge.

For those who would like to develop a more thorough understanding of the financial reports issued by their companies, many resource are available: the financial staff in their companies; the end-of-chapter references listed in this book; material such as the 4'/-page Merrill Lynch publication entitled, *How to Read a Financial Report*; and the briefer dissertation on this same subject contained in "A Road Map to Your Financial Report" by Frank C. Evans.

Definitions Applicable to the Balance Sheet

Balance Sheet. An accounting statement reflecting the financial status and stock wealth of an entity at a given point in time.

Current Assets. Cash and other assets that can be readily converted into cash, within a year.

Accounts Receivable. Money owed to the company by clients and customers, and others.

Inventory. Any supply of goods, may include items held for sale as finished goods, goods in production, raw materials, or supplies to be used in the production process.

As of December 31, Years 1 and 2 (in thousands)

	Year 2	Year 1
Current assets		
Cash and marketable securities	1,600	2,400
Accounts receivable	18,000	16,000
Inventory	28,800	26,000
Advances/prepaid expenses	800	800
Total current assets	49,200	45,200
Fixed assets		
Equipment at cost	34,000	30,000
Land and buildings	16,400	14,800
Total fixed assets	50,400	44,800
Less accumulated depreciation	19,600	16,000
Net fixed assets	30,800	28,800
Intangible assets	4,000	4,000
Total assets	84,000	78,000
Liabilities and Stockholder's Equity		
Current liabilities		
Notes payable	10,800	6,800
Accounts payable	13,200	14,400
Accrued expenses	2,000	1,600
Total current liabilities	26,000	22,800
Long-term debt	25,600	24,200
Total liabilities	51,600	47,000
Stockholder's equity		
Common stock, $10 par value, 1,000,000 shares issued and outstanding	10,000	10,000
Retained earnings	22,400	21,000
Total stockholders' equity	32,400	31,000
Total liabilities and stock-holder equity	84,000	78,000

EXHIBIT 1 *Sample consolidated balance sheet.*

Advances/Prepaid Expenses. Expenditures such as insurance premiums paid for policies that extend into coming years.

Fixed Assets. Resources with a life expectancy of more than one year.

Intangible Assets. An informal phrase for what is generally called "good will."

Current Liabilities. Sums that will become due and must be paid within a year.

As of December 31, Years 1 and 2 (in thousands, except earnings per share)	Year 2		Year 1	
Sales	152,000	100.0%	144,000	100.0%
Costs and expenses				
Cost of goods available for sale	159,600		147,040	
Less inventory, end of period	28,800		26,000	
Net cost of goods sold	130,800	86.0%	121,040	84.1%
Other expenses				
Sales and marketing	2,920	1.9%	2,960	2.1%
General and administration	10,440	6.9%	11,880	8.3%
Depreciation	3,600	2.4%	3,200	2.2%
Interest expense	1,680	1.1%	1,320	0.9%
Total—other expenses	18,640	12.3%	19,360	1.8%
Total costs and expenses	149,440	98.3%	140,400	97.5%
Net operating income	2,560	1.7%	3,600	2.5%
Miscellaneous income	240	0.2%	240	0.2%
Taxable income	2,800	1.8%	3,840	2.7%
Income taxes—all	920	0.6%	1,280	0.9%
Net income	1,880	1.2%	2,560	1.8%
Earnings per share	$1.88		$2.56	

EXHIBIT 2 *Sample statement of income.*

Accounts Payable. Money owed by the company to others.

Accrued Expenses. May include expenditures such as salaries and wages earned but not yet paid, taxes, accrued vacation.

Long-term Debt. Obligations that will become due after a year, e.g., bonds, mortgages, etc.

Par Value. The value assigned to authorized common stock for legal reasons; it bears no relationship to the market value of a share of stock.

Stockholders' Equity. The difference between the assets of an entity and its liabilities. Owner's equity is what belongs to the owners.

Definitions Applicable to the Income Statement

Income Statement. An accounting statement reflecting the earnings and expenses of an entity over time, usually year to year.

Costs of Goods Available for Sale. Includes all costs directly related to producing the product or providing a service. Costs for sales and marketing, and administration and general expenses are not included here.

Inventory. As described for the balance sheet.

Depreciation. The allocation of a portion of the acquisition cost of fixed assets during each of the time periods in which the assets are utilized.

Earnings per Share. Computed by dividing the net after-tax income available to common stockholders by the average number of common shares of stock outstanding for the period under consideration.

OTHER FINANCIAL STATEMENTS

Two other financial statements are typically found in annual reports. A *statement of changes in shareholder equity* indicates whether the net income was reinvested in the company, or distributed as earnings per share; or if the net income was negative, it tells how equity was affected. A *statement of cash Flow* sets forth in detail the inflow and outflow of cash and concludes with a figure that establishes an increase or decrease in cash at year-end.

PERFORMANCE RATIOS

In the texts cited here, considerable space is given to the many financial *performance ratios* that have evolved over time. The illustrations of performance measures and their definitions are for the most part very similar. Investors use them to determine the financial health of a company, and the quality of management. In discussions with investors and stockbrokers, executives may emphasize the goals they have established in relation to particular performance measures to obtain favorable treatment. These measures are called financial ratios. Very simply, a ratio is produced by dividing one number by another.

Performance ratios are important to safety practitioners who want to know what measures are currently applicable in their companies and how the financial results achieved at a given time may affect the responses to the safety expenditures they propose. Having such knowledge may also affect the timing of those proposals. Performance measures in use may differ from company to company and may be transitory over time within a company. Measures applicable in a given year may be replaced subsequently by others.

As an example, at a prominent, asset-rich company, the current performance micrometrics, which have been accepted by investors as being presently relative to the company achieving the financial goals set for a stated return on net assets, are:

1. Inventory turnover
2. Facility consolidation (percent of utilization)
3. Overhead cost management
4. Supplier-base consolidation

These performance indicators, and their definitions, are specific to this company. They are considered significant to the company in relation to its current business status and needs.

Generally, financial texts describe four principal types of financial ratios: liquidity ratios; leverage ratios; activity ratios; and profitability ratios. In this text, we only address profitability ratios, since they are most prominent in financial discussions.

PROFITABILITY RATIOS

Profit margin, *return on investment*, and *return on equity* are commonly used terms in the language of finance. Profit margin, simply stated, is the ratio of the net profit before taxes to revenue, or sales, expressed as a percentage.

What is developed in computing profitability ratios is an indication of the profit attained and how it relates to the revenue generated (margin), or to the resources employed (return on investment), or investor equity (return on equity). Here are the three formulae:

$$\text{Profit margin on sales} = \frac{\text{net profit before taxes}}{\text{sales}}$$

$$\text{Return on investment} = \frac{\text{net profit after taxes}}{\text{total assets}}$$

$$\text{Return on equity} = \frac{\text{net profit after taxes}}{\text{equity}}$$

Return on equity (ROE) differs from return on investment (ROI). "Investment" includes all the assets applied in the business, including borrowed money, in the short or long term. "Equity" includes shareholder equity only. For example: A company has an after-tax profit of $500,000 and its total investment is $5,000,000. Using the ROI formula—net profit after taxes divided by total assets—we determine the return to be 0.10, or 10%.

Assume that of the $5,000,000 investment, $3,000,000 is borrowed money (debt). If we use the ROE formula—net profit after taxes divided by equity ($500,000 divided by $2,000,000)—the result is 0.25, or 25%. If that number seems high, it should be understood that some companies produce returns on equity better than this.

The use of borrowed money, recorded as debt in financial statements, to finance a business "leverages" the stockholder investment to attain higher profits. So, debt is not altogether bad if revenues are sufficient to pay for the cost of the borrowed money and still produce an adequate return on equity. Almost all companies borrow money to underwrite capital purchases or purchase new assets, thereby "leveraging" profit potential. Some entrepreneurs have been very successful in building businesses and fortunes on borrowed money.

Safety practitioners need to understand that great differences exist in the expected margins, return on investment, and return on equity in different businesses—especially margins. As examples: A margin of 0.5% in supermarkets is not unusual and may be acceptable; a margin of 30% may well be expected and exceeded in some electronics businesses. Whatever management's expectations are, safety practitioners should be aware of them.

FINANCIAL DECISION MAKING TECHNIQUES

In "Using Financial Principles in the Safety Profession," Hansen, Grotewold, and Harley posit the following:

> To be understood, the safety manager must speak a language that upper management understands—dollars and cents. To cogently persuade upper management to approve an expenditure, one must illustrate why spending money now will save money later (37).

The authors then proceed to describe several "engineering economy" principles, among which are present worth, rate of return, and cost-benefit analysis.

Three other recent articles in different publications reflect a growing interest in safety practitioners communicating in the language of management:

- "Ergonomic Benefit/Cost Analysis," by Bruce K. Lyon, *Professional Safety*, March 1997 (33)
- "IH Professionals Learn to Apply Financial Concepts to Health and Safety," by Christopher O'Leary, *The Synergist*, June/July 1996 (17)
- "Cost-Benefit Analysis of Ergonomic Programs," by Maurice Oxenburg, *The American Industrial Hygiene Journal*, February 1997 (151)

The author was able to locate only one on-going course that specifically relates financial assessment techniques to safety. Organizational Resource Counselors, Arthur Andersen LLP, and 15 industry leaders have developed what they call the "Return on Health and Safety Investment (ROHSI)" process. Its promotional material commences with these observations: "Faced with some tough questions these days? Need to demonstrate value to your company? Increasingly, managers must demonstrate that health and safety investments create competitiveness in their business."

Its developers further claim that ROHSI should be used when evaluating capital and investment decisions, in the decision making to determine what health and safety investments ought to be made, and in identifying which health and safety investments create the greatest value to an organization and how that value can be demonstrated.

Literature for the ROHSI course indicates that discussions will cover, among other topics, net present value, return on investment, payback period, internal rate of return, and qualitative ranking of investment alternatives.

In the financial management literature reviewed by the author, discussions on the analytical methods mentioned by Hansen et al. and covered in the ROHSI material are extensive. Therefore, we will only briefly comment on the mathematical decision-making tools. They are cost-benefit analysis techniques.

COST-BENEFIT ANALYSIS

Cost-benefit analysis is at the core of management decision making. Further, the process is used by each of us. When we consider the outlay of money for a particular benefit, we are making cost-benefit decisions. Managers routinely use cost-benefit analytical techniques when considering a course of action and the alternatives available, especially for proposals involving capital expenditure.

It may be that the idea of performing cost-benefit analyses will not be easily accepted by safety practitioners who have become accustomed to presenting their proposals on risk avoidance, elimination, or control without a consideration of related costs, alternatives, or the risk reduction to be obtained.

Federal government agencies, such as OSHA and EPA, routinely conduct cost-benefit analyses. They are required to establish a balance between the costs to comply with a regulation and the benefits to be obtained, expressed in financial terms.

Cost-benefit analyses tools are a form of microeconomics and their use is far from an exact science. Determining the costs and the reality of the benefits to be obtained requires considerable judgment. Spiro offers these words of caution (20):

> Even if costs and benefits have been conceptually delineated to the analyst's satisfaction, representative estimates are difficult to develop. Furthermore, the validity of the data diminishes rapidly the further out in time the proposed action takes place. Yet, as any manager knows, this is a universal problem and decisions must be made in the face of uncertainty.

Even though the use of cost-benefit analysis techniques is as much art as science, these techniques are the methods management personnel apply in assessing expenditure proposals. We will now briefly comment on each of the techniques previously mentioned, with an introduction concerning the time value of money.

THE TIME VALUE OF MONEY

Donnahoe devotes an entire chapter to the "power of compound interest" (35). His purpose is to highlight the understanding that the future value of a current investment is considered in the application of project analytical methods, rather than the cost of the initial investment. As an example, for a $1,000 investment for a period of 5 years at 6% interest, compounded, the principal increases to $1,338.

Why make so much of the *time value of money*? As Friedlob and Plewa point out:

> A dollar received today is worth more than a dollar to be received at some time in the future. Money received today is worth more because it can be invested and today's dollars will yield more than the invested amount over a period of years (159).

Future Value

To obtain the *future value* of an investment, a formula is used in which the principal, the amount invested, is multiplied by the interest rate, expressed as a decimal plus 1, for the number of years being considered. An example follows:

Principal: $1,000
Interest rate: 8%
Number of years: 5
Future value = $1,000 × 1.08 × 1.08 × 1.08 × 1.08 × 1.08
Future value = $1,469.33

Such lengthy computations are not actually made, since interest tables are available, relating the number of years being considered to the rate selected. They simplify the exercise greatly. Also, calculators and computer programs sold in recent years can make these same computations. Table 1 gives an indication of the significance of compounding for an investment. The base used is $1.

TABLE 1 Future Value of $1 due in *N* Years

	Interest Rate (%)				
No. of Years	6	7	8	9	15
1	$1.060	$1.070	$1.080	$1.090	$1.150
2	1.124	1.145	1.166	1.188	1.323
3	1.191	1.225	1.260	1.295	1.521
4	1.262	1.311	1.360	1.534	1.749
5	1.338	1.403	1.469	1.539	2.011

Net Present Value

Determining *net present value* is a very important and commonly used management decision-making technique. In determining the net present value of an investment, the concept of compound interest is applied in reverse, e.g., the present value of $1,469.33, as computed at an interest rate of 8% in the preceding example for Future Value, to be received 5 periods later is $1,000. Friedlob and Plewa explain:

> Net present value is a discounted cash flow technique where all expected cash inflows and outflows are discounted to the present point in time, using a preselected discount rate. The present value of inflows is added together, and the initial outlay (and any other subsequent outflows) is subtracted (185).

What is discounted cash flow? Referring again to Friedlob and Plewa:

> Discounted cash flow refers to the present value, at a point in time, of a stream of receipts or payments to be received or paid in the future. The receipts or payments pattern is called an annuity (160).

Two examples of net present value computations are given here. In the first example, the mechanism, i.e., the thought process, on which the computation is based is demonstrated. In the second example, the intent is to show how the process can be simplified, using an annuity table. In case 1, computations made use discount rates; in case 2, a multiplier for an expected rate of return, taken from an annuity table, is used. The discount rates used in the first computation are shown in Table 2.

TABLE 2 Present Value of $1 due in *N* Periods

	Interest Rate (%)				
No. of Years	6	7	8	9	15
1	0.943	0.935	0.926	0.917	0.870
2	0.890	0.873	0.857	0.842	0.756
3	0.840	0.816	0.794	0.772	0.658
4	0.792	0.763	0.735	0.708	0.572
5	0.747	0.713	0.681	0.650	0.497

Case 1

A company is considering the replacement of aging equipment at a cost of $680,000. Production personnel compute after-tax labor cost savings at an annual average of $178,000. For the expected life of the equipment, 5 years, the discount rate used in the net present value calculations is 7%. Thus, from Table 2, we obtain

Present value, year 1:	$178,000 × 0.935 =	$166,430
year 2:	178,000 × 0.873 =	155,394
year 3:	178,000 × 0.816 =	145,248
year 4:	178,000 × 0.763 =	135,814
year 5:	178,000 × 0.713 =	126,914
	Total	$729,800

The present value of the inflows, $729,800, exceeds the present value of the outflow, the initial cost of $680,000. Since the net present value is positive, it is likely that the investment would be made. If the discount rate used was higher, the net present value would be lower. For example, at a 15% discount rate, the net present value would be $596,834, which is less than the investment cost. Very few projects are approved when the net present value is negative, unless there are other compelling circumstances.

The previous computation shows, fundamentally, how the system works. In actual practice, one would use an annuity table to select the multiplier for an expected rate of return of 7% for 5 years. That multiplier is 4.1: The annual benefits, $178,000, multiplied by 4.1 equals $729,000. This is the same net present value as shown above.

Case 2

Assume that a company plans to invest $500,000 in new equipment for which it expects annual gains of $200,000, after a careful review of all considerations. A rate of return of 15% over 5 years is to be achieved. In the computation, the multiplier used with respect to a 15% return on investment is the sum of the present value indicators for 15% shown in Table 2, for 5 years—thus, 3.353. Again, such summations need not be made since present value tables are available, giving the expected rates of return for a number of years. Thus,

Investment cost:	$500,000
Annual gains:	200,000
Return on investment multiplier:	3.353
Net present value = 200,000 × 3.353 =	$670, 600

This project would be worthwhile, since the net present value is positive, and a viable candidate for funding.

It should be kept in mind that the two examples of net present value just shown were intentionally simplistic. They were offered for the purposes of illustration only.

PAYBACK METHOD OF PROJECT EVALUATION

Rather obviously, the *payback period* is the length of time required to pay back the initial investment through the cash flow generated by the investment. Note that the consideration here is cash flow: in this instance, the sum of the after-tax income and the allowance for depreciation on the investment. For example: A company plans to make an investment of $1,500,000 for which the annual, after-tax flow is estimated to be $325,000:

$$\frac{\text{Original investment}}{\text{After-tax cash flow per year}} = \text{payback period}$$

$$\frac{\$1{,}500{,}000}{325{,}000} = 4.62 \text{ years}$$

If 4.62 years is within the payback time acceptable to the company, the investment would be favorably considered.

INTERNAL RATE OF RETURN METHOD OF PROJECT EVALUATION

Unfortunately, the literature on using the *internal rate of return method* in determining the acceptability of a proposed project is not uniform. More than one author cautions about the reliability of the method in determining project acceptance, principally because of its inability to compensate for variations in income over time. Many experts prefer the *net present value method*, in which such variations can be accommodated. Nevertheless, the internal rate of return analytical technique is often used.

The purpose of the internal rate of return method is to determine whether the financial gains to be made through an investment will equal or exceed the rate of return that a company has established as its minimum. That rate of return may be called the *hurdle rate*. If the computed internal rate of return is higher than the established minimum, the project would probably receive favorable consideration.

Calculator and computer programs are available to make the necessary calculations. Annuity charts also exist for the simple calculations, an example of which follows.

A company plans to spend $100,000 for a new system, for which the annual labor savings are estimated to be $30,500 per year, over 5 years. After 5 years, it is assumed the system will have no monetary value. Return on the investment needs to be at least 15%. The formula is as follows:

$$\frac{\text{Investment}}{\text{Savings}} = \text{Annuity factor}$$

$$\frac{\$100{,}000}{30{,}500} = 3.279$$

Referring to the annuity charts, we determine that the expected return for a 5-year term having a factor of 3.279 is 16%. Since the expected return exceeds the hurdle rate, the established minimum of 15%, the project looks favorable. Now, assume the same investment and an annual savings of $23,000:

$$\frac{\$100{,}000}{26{,}300} = 3.802$$

In the annuity table, a factor of 3.802 for a 5-year term indicates a 10% return, which would not be acceptable.

BREAK-EVEN ANALYSIS

Break-even analysis is a profit planning technique that considers fixed costs, variable costs, and the income from sales necessary to arrive at a financial "break-even" point, at which point total income equals total costs. It is expected that the profit produced by an operation would continue to grow beyond that point. A break-even point may be expressed in units of production or dollars of sales. Also, the profit expected per unit of sales is factored into the formulae given in some of the literature.

Fixed costs are those that remain unchanged, mostly but not entirely, regardless of the amount of the production or service output. They include expenditures such as management salaries, rent, some taxes, some insurance premiums, and depreciation. Variable costs will increase as the production or units of service provided grow, such as direct labor, materials, worker's compensation premiums, and public utility uses.

The break-even point is a factor of the fixed and variable costs, and the potential sales price. Methods depicted in the literature showing the convergence of costs and revenues to reach a break-even point differ somewhat, but they achieve the same result. Variations of the chart in Figure 1 appear in several texts. Some of the charts are elaborate: In crafting Figure 1, simplicity was our aim.

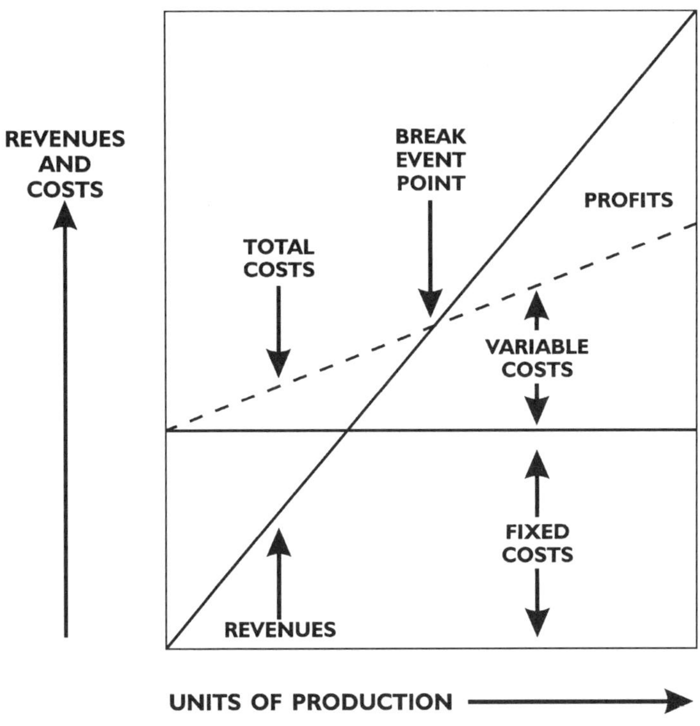

FIGURE 1 *Break-even chart.*

In the charts, a horizontal fixed cost line is illustrated, with an indicator of continuously rising variable costs. The summation of both cost categories is shown as an increasing total cost line as the units of production and revenues grow. The break-even point is where the total revenue line crosses the total cost line. Profit begins and increases beyond that point.

Although the formula to develop the break-even point may seem daunting, knowledge of only simple algebra is needed to complete the computation. To express the break-even point in units of sale, a computation applying the formula below would be made, using the following variables:

P = expected sales price per unit

F = fixed cost per unit

V = selected average of the variable costs per unit

Q = quantity produced and sold

$$P \times Q = F + (V \times Q)$$

$$(P \times Q) - (V \times Q) = F$$

$$Q \times (P - V) = F$$

$$Q = \frac{F}{P - V}$$

Assume that fixed costs are \$400,000, the sales price per unit is \$50, and variable costs per unit are \$18:

$$Q = \frac{400{,}000}{50 - 18}$$

$$Q = 12{,}500 \text{ units} = \text{the break-even point}$$

To determine the number of units that must be sold to achieve the desired profit, the profit is added to the fixed costs. Then, the formula becomes

$$Q \text{ (to include profit)} = \frac{\text{profit} + F}{P - V}$$

Assume that the desired profit from this operation is \$150,000. Then, the calculation is as follows:

$$Q = \frac{150{,}000 + 400{,}000}{50 - 18} = 18{,}750 \text{ units}$$

Andrew Sorine, in his article entitled, "Accounting for Safety" (53), expresses the view that safety practitioners should have a good understanding of break-even analysis methods and be aware of how such computations affect management decision making. The literature on this subject is extensive.

CONTRIBUTION MARGIN

The *contribution margin* is commonly defined as the difference between the amount of sales and variable expenses. Contribution margin computations are made for individual products or services, or for operating divisions or units. In the literature, the contribution margin is eventually related to profitability, which requires a consideration of all costs. An example of contribution margins for two operating divisions is presented in Table 3.

TABLE 3 Example: Contribution Margin Comparisons

	Division A	Division B
Sales	$1,000,000	$1,500,000
Variable costs	660,000	800,000
Contribution margin	340,000	700,000
Fixed costs	350,000	420,000
Pretax profit	(10,000)	280,000

TABLE 4 Contribution Margin Related to Units of Goods Sold

No. of units sold	500,000
Price per unit	$20
Total revenue	$10,000,000
Variable costs	$8
Contribution margin	$12
Total contribution	$6,000,000
Fixed costs per unit	$10
Total fixed costs	$5,000,000
Pretax revenue	$1,000,000

Some serious decision making would be necessary concerning the entity producing a pretax loss. When the contribution margin is related to units of goods sold or services provided, it is expressed as the difference between the unit price and the variable costs per unit. An example of this is given in Table 4.

In the process of determining the contribution margin, the proper allocation of expenses is critical. Allocating variable costs can be done with considerable accuracy and usually with little controversy. The allocation of fixed costs is not so easily done. Business unit managers often dispute fixed cost allocations, which are difficult to understand, particularly when their financial well-being is affected.

REAL-WORLD APPLICATIONS

As financial management is not an exact science, neither is the decision making about risk avoidance or reduction. Safety practitioners need to understand the implications of fear or dread, loss of cash flow, or employee concerns in the decision making, and when and how to take them into account. Consider these two examples.

Case A

An environmental, health, and safety manager in an operation using a minor amount of chemicals in producing a product has to address the problem of environmental control equipment that operates erratically. Occasionally, a few gallons of a chemical spills into the adjacent river. Working with engineering personnel, it is determined that there are mechanical and circuitry design problems. The cost of correcting them is estimated at $10,000. It is further determined that a spill of 1,000 gallons of the chemical would probably occur once in 5 years. At $0.95 a gallon, the product loss would be $950. At first, management declines the safety expenditure request, not finding much logic in spending $10,000 to protect against a loss of $950 that might occur once every 5 years. Then, the amount of the product loss and damage increases to $1,000, the additional $50 representing the exaggerated value of the estimated 25 fish that will perish as a result of the spill. Just imagine the discussions that take place among management personnel who would have to respond to inquiries from governmental agencies and the print and broadcast news media. After a consideration of such consequences, the expenditure is approved.

Case B

In a chemical operation, outside consultants are retained to work with staff engineers in determining the worst-case safety scenario, one that might severely damage operations. It was agreed that a pipe and valve failure which would release a large quantity of combustible liquid was such a catastrophe. The consultants and internal engineers estimated the direct property damage loss at $400,000,000, which was 20% of the location's replacement cost. It was further estimated that there would be 4 fatalities and 12 serious injuries if such a failure occurred.

Additional deluge systems and fire pumps were recommended, which would cost $600,000. Because of exceptionally high insurance retentions, no insurance cost savings would derive from these installations.

The probability of a pipe and valve failure became a point of contention. At first, management declined approval of the recommended safety expenditure. There were 10 similar facilities in the world, and it was determined that, among them, a major loss occurred once every 30 years. That statistic translated into the probability of a pipe and valve failure once every 300 operating years. When the corporate financial implications were more carefully considered, however, the decision to reject the safety proposal was reversed. All 10 worldwide operations operated at nearly full capacity, which meant that the company would likely not be able to buy enough replacement product to fill customer orders. The repair time for the damage was estimated at a minimum of 18 months, but more likely 24 months.

More important, the cash flow from this location was vital to both the continuation of corporate dividend payments, and to the payments due on the highly leveraged debt. Without the continuing cash flow, the company could face bankruptcy.

These examples suggest how safety practitioners can become involved in executive decision making, going beyond the mere presentation of a list of safety recommendations.

CONCLUSION

This chapter is a primer, only a primer. Its purpose is to provide introductory knowledge on the basics of financial management. Some readers, particularly after determining the analytical methods and the performance ratios that are important in their companies, will probably want more information about them. It is suggested that they then initiate discussions with the financial management personnel in their organizations for educational purposes. Most financial departments have programs to calculate economic returns and personnel who know what the decision makers expect to see in expenditure requests. Safety practitioners seeking their counsel might receive favorable and encouraging responses to their inquiries.

They should keep in mind, as one learned in the field recently said, that financial management is somewhat of an art. In all the analytical systems mentioned in this chapter, the numerics to be entered into computations will involve many judgments. Often, estimates will be necessary since hard data will not be available, as is often the case in risk assessment. Also, safety practitioners may find that financial management personnel within their companies will provide definitions for the analytical systems used that differ from those described here. Whatever their definitions, they are the definitions that are operable.

Although this chapter was written in response to safety practitioners' requests for a better understanding of the basics of financial management, it would be inappropriate not to speak of other ways by which safety practitioners may be perceived as providing value. Understanding financial basics surely contributes to one's being viewed as "on board" and a "part of the management team." But, the substance of what a safety practitioner does also creates impressions of value, favorable or unfavorable.

REFERENCES

ASSE Risk Management/Insurance Division. "1998 White Paper, Loss Control 2000." *ASSE Risk Management/Insurance Newsletter*, Winter 1998–99.

Bird, Frank E., Jr. and Ray J. Davies. *Safety and the Bottom Line*. Loganville, GA: Febco, Inc., 1996.

Deming, W. Edwards. *Out of the Crisis*. Cambridge, MA: Center for Advanced Engineering Study, Massachusetts Institute of Technology, 1986.

Donnahoe, Alan S. *What Every Manager Should Know About Financial Analysis*. New York: Simon and Schuster, 1989.

Drucker, Peter F. *Post-Capitalist Society*. New York: HarperCollins Publishers, 1993.

Environmental, Health and Safety Report 1998. New Brunswick, NJ: Johnson & Johnson.

Evans, Frank C. "A Road Map to Your Financial Report." *Management Review*, Oct. 1993.

Friedlob, George T. and Franklin J. Plewa, Jr. *Understanding Return on Investment*. New York: John Wiley & Sons, 1996.

Hansen, Mark D., Henry W. Grotewold, and Richard M. Harley. "Using Financial Principles in the Safety Profession." *Professional Safety*, June 1997.

How to Read a Financial Report. New Brunswick, NJ: Merrill Lynch, 1997.

Lyon, Bruce K. "Ergonomic Benefit/Cost Analysis: Communicating the Value of Enhancements." *Professional Safety*, March 1997.

Manuele, Fred A. *On The Practice Of Safety*. New York: John Wiley & Sons, 1997.

Muckian, Michael. *The Complete Idiot's Guide to Finance and Accounting*. New York: Alpha Books (Macmillan), 1998.

O'Leary, Christopher. "IH Professionals Learn to Apply Financial Concepts to Health and Safety." *The Synergist*, June/July 1996.

Oxenburgh, Maurice. "Cost-Benefit Analysis of Ergonomic Programs." *American Industrial Hygiene Journal*, Vol. 58, Feb. 1997.

Shim, Jae K. and Joel G. Siegel. *Financial Management for Nonprofits*. New York: McGraw-Hill, 1997.

Sorine, Andrew. "Accounting for Safety." *Occupational Hazards*, Sept. 1994.

Spiro, Herbert T. *Finance for the Non-Financial Manager*, 4th ed. New York: John Wiley & Sons, 1996.

Van Horne, James C. and John M. Wachowicz, Jr. *Fundamentals of Financial Management*, 10th ed. Upper Saddle River, NJ: Prentice-Hall, 1998.

Weston, J. Fred, Scott Besly, and Eugene F. Brigham. *Essentials of Managerial Finance*. Fort Worth, TX: Dryden Press, 1996.

2

Establishing Your Value: Communicating Incident Statistics in Financial Terms

INTRODUCTION

Financial reports to management on hazards-related incidents must establish the value of the contributions made by safety practitioners to an entity's financial well-being. This chapter provides guidance on preparing such reports.

Safety practitioners accept as a given that superior success in safety, long term, can be achieved only when safety is a significant value within an organization's culture. Also, it must be understood that trying to improve safety results principally for money savings without taking steps to achieve the necessary changes in culture is a questionable endeavor. Nevertheless, management relies on the statistical outcomes of safety-related initiatives as performance measures.

In developing and presenting hazards-related statistics for management consideration, the following premises should be considered:

1. Finance is the language of management, and reports with financial implications are more likely to capture management's attention.
2. Management usually sets goals for improvement based on previous results.
3. Exhibits developed should be plausible and supportable with sound logic.
4. Financial exhibits of direct and indirect costs can be very impressive, but it is important to be realistic in compiling them.
5. As far as practical, studies and exhibits should include all hazards-related

incidents: employee injuries, property damage, fires, machinery breakdown, public liability, auto incidents, product liability, etcetera.

6. Safety results are determined by an organization's culture, and an organization's culture changes slowly: Unless significant measures are taken impacting on an entity's culture, its financial history is a sure predictor of the future.
7. When the sample base is large enough (e.g., the number of hours worked), the historical OSHA incident record has a good degree of confidence as an indicator of the quality of safety management in place, and for statistical predictions of the future.
8. Statistical quality-control methods, particularly simple control charts, can advantageously and plausibly display the trends in incident rates and financial data.
9. For the company that purchases insurance, its experience modifications (for all insurance categories) are financial indicators of the quality of safety in place, and of financial trending.

This chapter is divided into three parts. Part 1 gives examples of statistical analyses and presentations with financial implications made by safety practitioners. Part 2 is devoted to computing the financial implications of direct and indirect hazards-related incident costs. Part 3 discusses other aspects of performance measures and their financial implications.

Part 1
Statistical Analyses and Presentations With Financial Implications

Real-world examples of incident analyses made by safety practitioners follow. Their presentation influenced decision makers to conclude that the work of safety personnel provides financial value, or that doing what they propose would result in financial gain.

Sometimes, and this is important, analyses made by safety practitioners were expressed in operational terms, not dollars. Nevertheless, decision makers could easily translate those terms into financial implications.

At one motivational presentation, a sales professional said that almost everyone has a "hot button" which, when hit, produces a favorable response. Timeliness is important in determining what management's hot button may be at a given time: e.g., pressure on earnings, the need for improved productivity or lower costs, or superior quality.

- A safety director obtained numbers that management agreed were reasonable estimates of annual workers compensation costs. She determined that the company's OSHA recordable and lost workday case rates were about one-fifth that of the industry average. Through a simple extrapolation, she computed that the company's annual workers compensation costs would be $40,000,000 higher if its OSHA rates were at the industry average. That got the attention of management.

- For a company engaged in a highly competitive market, adding a nickel to earnings per share is a real accomplishment. The safety director and others computed workers compensation costs to be $0.45 an hour. That data, along with proposals for an extended safety initiative, were presented to the CEO. He jested that he would kill for a 20% reduction in those costs.

  Why did the CEO have such a response? A savings of $0.09 an hour, for 2,000 hours worked, translates to $180 per employee, per year. For 30,000 employees, that becomes $5,400,000 annually. With 40,000,000 shares of common stock outstanding, a 20% reduction in workers compensation costs improves earnings by $0.135 per share. An improvement of that magnitude would be a significant accomplishment.

- In a very large company, the safety director recognized that there were culture problems to be overcome. Senior executives did not follow through on the policies they publicly espoused. Although they delegated responsibility for safety to their staffs, their personal involvement was limited. More important, real accountability was absent. Efforts to influence the senior staff to recognize the benefits of reducing incidents and their related costs were not successful—until an independent study indicated that the hours of labor needed per unit of production were considerably higher than those of competitors. That finding became the focus of management concern.

  The safety director, sensing an opportunity prepared a report to management based on statistical quality-control principles. The report indicated that the consistent trends of OSHA recordable incidents and lost workday cases indicated that the safety management system was in control, although at an unacceptable level. Further, the report indicated that if no substantive, operational change was made in the safety management system, the past predicted the future and management should expect a continuation of the previous experience, with slightly higher costs.

  One of the projections made in the report was that in the following year, about 1,350 employees would be away from work at some time because of on-the-job injuries. Obviously, they would not be producing product. Management was disturbed by that statement and consequently became more open to safety suggestions. In addition, it was tactfully suggested that they were part of the problem, since safety was not really a value in their company. Management's hot button had been hit, and real changes ensued as a result.

- In a company with 2,600 employees, the workers compensation experience rating modification was 1.54 when the safety practitioner was hired. This meant that the employer was paying 54% more than the average paid by other employers in the same business. Over 5 years, the experience modification was reduced to 0.68, and annual premium costs dropped from $1,221,400 to $539,200. Net dollar annual savings were $682,200. Reports prepared by the safety practitioner charted the financial improvement.

- In an electronics assembly operation, back injuries and other strains and sprains were frequent and costly. After analyzing the incident history, the safety practitioner, in her discussions with manufacturing engineers observed that the areas in which injuries were occurring were also problem areas for productivity and unit costs. The safety practitioner suggested an ergonomics study, which revealed that both the injury and productivity problems had been designed into the work process. The physical layout and the work methods overstressed a large part of the employee population, and efficiency was seriously hampered as a result.

  Jointly, the manufacturing engineers and the safety practitioner recommended new equipment that addressed the ergonomics problems. Its cost was estimated at $275,000. After a year, the safety professional produced a report showing that, remarkably, the investment had been recovered through injury savings alone during that time period. Since manufacturing engineers had also computed notable productivity gains, the safety practitioner received a "functional promotion" in management's view because they recognized that her work had favorably influenced company operations beyond safety.

> Anecdotal data describing productivity, cost, and quality improvements that result from successful ergonomics applications initiated for safety purposes are voluminous. Ergonomics is a fertile field in which safety practitioners, with others, can produce impressive reports indicating financial gains, reports that thus establish their value.

- For years, a safety practitioner could not obtain an acknowledgment from the vice-president for sales that the operating rules for the use of company automobiles needed improvement, even though it was widely recognized that accident rates were high. Typical explanations were given, such as the pressure under which salespersons normally operated and their need to be at certain places at certain times. Another consideration was the customary entertainment of clients, where alcohol consumption was the norm.

  Within the company, all of the few fatalities experienced over a period of many years involved salespersons. Disabling injuries resulting from auto accidents were also prevalent among this segment of the workforce. Nevertheless, the related cost data that had been compiled did not influence the vice-president for sales.

  The safety director became aware that one of the important performance measures applied to salespersons was the average number of client calls made per day. The premise was that there was a direct relationship between the number of sales calls made and the company's overall revenues.

In his annual report, the safety director included an estimate of the number of sales calls not made because of auto accidents. The chairman of the board of directors read that report. Sales calls not made in his mind translated into revenues not produced. And, a minor explosion resulted.

Prompting sales executives to give the deserved attention to passenger car safety is not easy. In this case, a communication in terms that management understood got it done.

Part 2
Direct and Indirect Costs

Although communications on the direct and indirect costs of hazards-related incidents can be attention-getting, dramatic, and convincing, such exhibits should be plausible, realistic, supported with good data, and economically obtainable.

Although the literature on direct and indirect costs is extensive, it is also somewhat baffling. In *Safety & the Bottom Line*, a 1996 publication (272) Bird and Davies offer an interesting finding after a search of all available literature: Well over 90 published articles or books discussing accident costs have been written, since the originator of the concept of direct and indirect costs, H. W. Heinrich, went to press in 1931. It is not the intent of the author to evaluate everything that has been written on direct and indirect costs. Rather, we will focus our attention on H. W. Heinrich's premises; the cost-estimating system developed by Rollin H. Simonds, which was adopted by the National Safety Council; the writings of Frank E. Bird, Jr.; a study pertaining to the construction industry; and a recent development by OSHA.

A great variety of terms appear in the literature on the costs of hazards-related employee incidents: direct and indirect (incidental, hidden), insured and uninsured, ledger and nonledger, visible and invisible, tangible and intangible, objective and subjective. Mostly, the definitions of these terms are close to what have come to be understood as direct and indirect costs, which the author will explore here.

HEINRICH'S PREMISES

H. W. Heinrich's discussion of direct and indirect costs in *Industrial Accident Prevention: A Scientific Approach* was a historical first and is much referenced in the literature. What follows are excerpts from the 1st edition of his book (published in 1931), giving the results of his 1926 research and analysis.

> Hidden Accident Costs. To (business) executives is now offered the incidental, or more correctly, the hidden or indirect, but none the less real, employer cost of accidents, as indisputable evidence of the need for recognizing accident prevention as an essential in sound business management (17).

The many and varying estimates of the annual cost of industrial accidents are stated in terms of millions of dollars and are usually based on the lost time of the injured worker. This is largely an employer loss, inasmuch as the employee is partially compensated; but it is far from being all of the cost to the employer. The remaining additional, and so-called "incidental," cost has been found by analysis to be *four times as great* as compensation and medical payments. Expressed in another way, compensation payments constitute only one-fifth of the total employer accident cost. The accuracy of this estimate has been demonstrated by application to scores of specific plants. The Associated General Contractors of America state that the cost is probably higher in construction work (17).

It is not contended that the four-to-one proportion holds true for *every* industrial accident, nor *every* individual plant, and it is granted that in nation-wide application the ratio may vary; yet it has already been tested sufficiently to provide approximate confirmation (17).

Factors in the Hidden Cost, to Employers, of Accidents to Employees. The calculations from which this four-to-one ratio was derived were based on the factors that follow. Compensation and liability claims, medical and hospital cost, insurance premiums, and cost of lost time except when actually paid by the employer without reimbursement are excluded:

1. Cost of lost time of injured employee.
2. Cost of time lost by other employees who stop work
   a. Out of curiosity.
   b. Out of sympathy.
   c. To assist injured employee.
   d. For other reasons.
3. Cost of time lost by foremen, supervisors, or other executives as follows:
   a. Assisting injured employee.
   b. Investigating the cause of the accident.
   c. Arranging for the injured employee's production to be continued by some other employee.
   d. Selecting, training, or breaking in a new employee to replace the injured employee.
   e. Preparing state accident reports, or attending hearings before state officials.
4. Cost of time spent on the case by first-aid attendant and hospital department staff, when not paid for by the insurance carrier.
5. Cost due to injury to the machine, tools, or other property or to the spoilage of material.
6. Incidental cost due to interference with production, failure to fill orders on time, loss of bonuses, payment of forfeits, and other similar causes.
7. Cost to employer under employee welfare and benefit systems.
8. Cost to employer in continuing the wages of the injured employee in full, after his return—even though the services of the employee (who is not yet fully recovered) may for a time be worth only about half of their normal value.

> 9. Cost due to the loss of profit on the injured employee's productivity, and on idle machines.
> 10. Cost of subsequent injuries that occur in consequence of the excitement or weakened morale due to the original accident.
> 11. Overhead cost per injured employee—the expense of light, heat, rent, and other such items, which continues while the injured employee is a non-producer (18).
>
> This list does not include all of the points that might well receive consideration, although it clearly outlines the vicious and seemingly endless cycle of events that follow in the train of accidents (17).
>
> Although 11 hidden accident-cost factors have been specifically listed in this chapter, it should be emphasized that, in calculating the ratio of four-to-one between hidden and direct costs, only a few of the factors were employed, these being the ones for which data were readily available. This fact, coupled with the omission of calculations for spectacular accidents that resulted in huge hidden costs, fortifies the argument (32).

Heinrich, although confident that his 4-to-1 ratio had been tested sufficiently to provide an approximate confirmation, cautioned that the proportion may not hold true for every individual plant and that in nationwide application the ratio may vary. Also, note that Heinrich indicated all possible indirect cost factors listed were not included in his cost computations and he used only those for which data were available.

This latter point is important: Data on unusual, costly, and spectacular events that resulted in huge hidden costs were omitted. If those events for which there were huge hidden costs had been included, the resulting ratio would be higher than 4 to 1. But, some recent work indicates that, more likely, the indirect costs of injuries resulting in severe injury would occur at a ratio lower than 4 to 1. We will have more to say about this later.

THE APPROACH OF THE NATIONAL SAFETY COUNCIL

In the 11th edition of the National Safety Council's *Accident Prevention Manual for Business and Industry: Administration & Programs*, there is a brief discussion of calculating insured and uninsured worker incident costs (191). There are major similarities and major differences to the Heinrich approach. Excerpts from the Council's text follow:

> **Estimating Incident Costs**
>
> This discussion concerns the elements of cost most likely to result from a work incident and presents a method whereby an organization can obtain an accurate estimate of the total costs of its work incidents. (This procedure for estimating costs was developed by Rollin H. Simonds, Ph.D., Professor, Michigan State University, under the direction of the Statistics Division, National Safety Council.)

> **Method for Estimating**
>
> To be of maximum usefulness, cost figures should represent as accurately as possible the experience of the company. It is not useful to have a fixed ratio of uninsured to insured costs representing many different organizations in many different industries. Estimated costs of incidents in general do not take into account differences in hazards from one industry to another or the more important differences in safety performance from one company to another.
>
> Because the distinctions between direct and indirect costs are difficult to maintain, they have been abandoned in favor of the more precise terms "insured" and "uninsured" costs. Using these data, a company can estimate its incident costs with reasonable accuracy.

Rollin Simonds developed his cost analysis system some 50 years ago. It was common, then, for even the largest companies to buy workers compensation insurance, and the terms "insured" and "uninsured" costs were appropriate. Presently, nearly all large employers are self-insured to the extent legally allowed. And, most entities not self-insured that are large enough to hire a safety practitioner are likely to buy their workers compensation insurance on a cost-plus, retrospective basis.

An associate in the insurance industry offers as an example of current trending a company that self-insures the first $50,000,000 for every aspect of hazards-related incident costs (e.g., workers compensation, fires, product liability). It is the author's view that, for the purposes of computing direct and indirect costs, the terms insured and uninsured costs have become obsolete.

Note the emphasis in the Council's text on trying to be accurate in cost estimates. Simonds' method of determining insured and uninsured costs is presented by the National Safety Council as a means of accurately establishing costs of incidents in an individual company.

Insured costs, so the text says, can be determined from accounting records. Insured costs are not identical to direct costs as described in Heinrich's book. For example, the following instruction is given in the Council's text: "For companies not carrying their own insurance, it (insured cost) will be the amount of their insurance premiums plus deductibles paid."

Heinrich included only the cost of workers compensation claims paid in his computations. Administrative and other costs could be as high as 30% of the premium.

The National Safety Council proposes a pilot study to ascertain approximate averages of uninsured costs for each of the following four classes of incidents:

1. *Class 1.* Cases involving lost workdays: days away from work or days of restricted work activity.
2. *Class 2.* Medical treatment cases requiring the attention of a physician outside the facility.
3. *Class 3.* Medical treatment cases requiring only first aid or treatment at a local dispensary and resulting in property damage of less than $100 or loss of less than 8 hours in work time.

4. *Class 4*. Incidents that either cause no injury or cause minor injury not requiring the attention of a physician, and that result in property damage of $100 or more, or loss of 8 or more employee hours.

Ten categories of uninsured costs are listed in the Council's text. They bear considerable similarity to the hidden costs developed by Heinrich, but they are not identical. Also, what would be included in the uninsured category is somewhat restricted in relation to the Heinrich method.

In Grimaldi and Simonds' *Safety Management*, 5th edition, Chapter 6, is entitled, "Cost Analysis" (208). It contains a six-page discourse on the "Invalid and Restricted Items" that should not be included in computing uninsured costs. Here is a brief excerpt:

> Loss of profit of idle machines or workers is not generally a valid cost. When workers or machines are made idle, one of two things occurs. Either the production is eventually made up, or it is not. If it is made up, in the sense that over a long period of time no less goods are produced and sold than would have been had the "accident" not occurred, there is no loss of profit, apart from the increase in production costs.

To ascertain the uninsured costs, a pilot study should be conducted for a period of sufficient duration to include an adequate number of the types of incidents that can occur in the operation being studied. Its purpose is to develop average uninsured costs for the four classes of incidents used to establish current and future costs. This is a great idea. If such a pilot study can be made, reasonably accurate cost estimates could be produced. But, therein lies the rub: getting commitments from operations and administrative personnel for the time needed to complete the pilot study is difficult.

For those who seek further knowledge about computing insured and uninsured incident costs beyond the National Safety Council publication referenced here, the Grimaldi and Simonds text previously noted is a good resource. Appendix B illustrates the pilot study procedure recommended.

Recall that in the Heinrich study of hidden costs, calculations for spectacular accidents were omitted. Similarly, in the Simonds method, the costs of deaths, permanent total disabilities, and fires would not be included in the pilot study; their insured and uninsured costs would be calculated separately.

It is of interest that, in a doctoral dissertation at Northwestern University in 1948, Simonds stated that the ratio of insured to uninsured costs was probably close to 1 to 1. For many years, that ratio was used by the National Safety Council in the data presented in its *Accident Facts* publication. In the 1999 issue of *Injury Facts* (as *Accident Facts* was renamed), the Council presents a macro view of costs (51). It is illustrated in Table 1.

What is important here is the mix of costs included in the Council's summary: It differs considerably from other methods. Note that motor vehicle damage costs and fire losses are included in the National Safety Council computation.

TABLE 1 National Safety Council's Estimate of the Total Costs of Occupational Deaths and Injuries Injury Facts, 1999

Total cost in 1998	**$125.1 billion**
Wages and productivity losses	$62.9 billion
Medical costs	$19.9 billion
Administrative expenses	$25.6 billion
Money value of time lost by workers other than those with disabling injuries, who are directly involved or indirectly involved in injuries, and the cost of time required to investigate injuries, write up reports, etc.	$12.0 billion
Motor vehicle damage in work injuries	$2.2 billion
Fire losses	$2.5 billion

THE BIRD STUDIES

In his 1974 book, entitled, *Management Guide to Loss Control*, Frank Bird presented what he called the *iceberg theory of incident costs* (25). In an exhibit with the appearance of an iceberg and captioned, "The Real Costs of Accidents Can Be Measured and Controlled," Bird gave the ratios listed in Table 2.

TABLE 2 The Bird Ratios Management Guide to Loss Control, 1974

$1	Insured costs (medical and compensation)
$5 to $50	Uninsured property damage costs (building damage, tool and equipment damage, product and material damage, production delays and interruptions)
$1 to $3	Uninsured miscellaneous costs (items such as hiring and training, replacements, investigation time, etc.)

Depictions of the "iceberg" have appeared in many texts and articles, and its cost ratios are frequently repeated. Bird's finding that uninsured property damage costs are 5 to 50 times the insured costs is difficult to support universally. For uninsured miscellaneous costs, the 1-to-3 ratio Bird suggests is in the same league as the Heinrich 4 to 1 ratio and the 1 to 1 ratio originally produced by Simonds.

A CONSTRUCTION INDUSTRY STUDY

Under contract to The Business Round Table, an association of CEOs committed to improving public policy, Stanford University's Department of Civil Engineering issued a report in 1982 entitled, "Improving Construction Safety Performance." This study was initiated to provide guidance on reducing accident frequency and severity in the construction industry, and the attendant costs.

TABLE 3 Data from Stanford University Study on Improving Construction Safety Performance, Business Round Table Report

Range of Benefits Paid	Number of Cases	Average Ratio Indirect Costs/Benefits Paid
	No Lost Time	
$0–199	13	4.2
$200–399	7	5.1
$400 plus	4	5.0
	Lost Time	
$0–2,999	9	4.1
$3,000–4,999	8	1.6
$5,000–9,999	4	1.2
$10,000 plus	4	1.1

In the report, direct and indirect costs are discussed, their definitions differing from those of other methods. Nevertheless, the data presented deserve careful consideration. The data in Table 3 is taken from Stanford University Department of Civil Engineering Technical Report #260.

Some statisticians may argue that they would have liked a larger sample from which to draw conclusions. Nevertheless, the data in the Stanford University report are important. Whoever compiles direct and indirect incident cost data will find them illuminating, particularly the following paragraph.

> Analysis of the compiled data shows that the indirect cost ratio (or multiplier) is affected by a great many variables. These variables include the type of project, the diligence, of the investigation, the severity of the accident, how critical the affected project is to the construction contractor's client's activities, and more. So many variables are involved that it is not possible to provide a single multiplier for all construction industry accidents. Still, if the accident data are separated into two general groups, large and small claims, it can be seen that smaller accidents have a larger multiplier. Larger accidents, despite a smaller multiplier, require large benefits to be paid; so the magnitude of the indirect costs is still substantial.
>
> The varied multipliers in Table 3 are believed to be conservatively low; the total indirect costs are apt to be considerably higher.

Just how significant was the Stanford University study? OSHA developed a costing system based on it.

OSHA'S SAFETY PAYS PROGRAM

OSHA's Safety Pays program can be accessed through the World Wide Web at www.osha.org—Outreach—OSHA/Software Advisors. The literature describes the program's purpose as follows:

> OSHA's "SAFETY PAYS" program is interactive software developed by OSHA to assist employers in assessing the impact of occupational injuries and illnesses (with Lost Work Days) on their profitability. It uses a company's profit margin, the AVERAGE costs of an injury or illness, and an indirect cost multiplier to project the amount of sales a company would need to generate in order to cover those costs.

For each of 32 types of injuries and illnesses—from amputations to vision loss—the program will produce an average direct cost, average indirect cost, and the additional sales needed to cover indirect costs and total costs for various profit margins. If the user does not enter a profit margin, the system uses a default margin of 3%.

Printouts for the 32 injury and illness types were obtained and reviewed by the author. The findings are listed below. Keep in mind that the OSHA system only pertains to lost work-day cases.

- Safety Pays is a close adaptation of the construction industry data given in Table 3.
- Computations were made to determine the indirect to direct cost ratios used in each printout, and the results were sorted by the indirect ratio used.
- Dollar ranges for the direct benefit costs used are identical.
- The ratios of indirect to direct costs used also are the same, except for the lower direct cost category: OSHA used a 4.5 multiplier, rather than 4.1.
- The indirect to direct cost ratio used was the same for each of the injury or illness types when the direct cost was in the same dollar range.
- Figures entered under "Range of Average Direct Costs" are those shown in Table 3 of the construction industry study. The actual average direct costs for the data shown in printouts from the OSHA program closely match each of the four lost work-day dollar ranges in Table 3 here. They are provided in Table 4.

Since the OSHA program is clearly an adaptation of the construction industry study, it follows that the ratio of indirect to direct costs decreases as the direct costs increases. This is shown in Table 5.

It is the author's understanding that the average direct costs for injuries and illnesses in the OSHA study were compiled by an insurance company and that its report covered a four-year period. If cost data from the OSHA system are

TABLE 4 Actual Dollar Ranges of Direct Costs Taken from the OSHA Safety Pays Program

\$317–2,169
\$3,283–4,245
\$5,475–9,667
\$10,615–\$27,547

TABLE 5 Analysis of OSHA's Safety Pays Program

Range of Average Direct Costs	Number of Injury Type Categories	Ratio of Indirect to Direct Costs
$0–2,999	13	4.5
$3,000–4,999	3	1.6
$5,000–9,999	10	1.2
$10,000 plus	6	1.1

used, over time, inflationary multipliers would be appropriate. The OSHA program carries an August 1998 date.

The Safety Pays system adds to the body of knowledge on direct and indirect costs and should be seriously considered by anyone compiling direct and indirect incident costs.

OBSERVATIONS ON DIRECT AND INDIRECT COSTS

From the previous discussion of direct and indirect costs of worker injuries and illnesses, it must be concluded that the literature does not present a uniformly accepted computation method. In fact, the differences in the various systems are remarkable.

Unfortunately, many safety practitioners have adopted one specific ratio of indirect to direct costs, without the benefit of supporting data. They apply the ratio across a variety of business types, presenting it as a well-established concept that is universally applicable. Colleagues in large companies with a diversity of operations have indicated that, in communications to management, they use a single ratio of indirect to direct costs, and those ratios vary from 1 to 1 to 10 to 1. A plausibility problem thus results.

Ted S. Ferry made this comment in *Modern Accident Investigation and Analysis, an Executive Guide* on the direct and indirect costs of mishaps (4):

> Estimates of the indirect costs of a mishap range from once again as much as the direct costs to 200 times as much. A conservative ratio is 3:1: that is, the indirect costs run an average of three times the direct costs.

This is how Dan Petersen expressed concern over the use of a direct to indirect cost ratio without supporting data in his book *Techniques of Safety Management* (132):

> Although hidden costs are very real, they are very difficult to demonstrate. To say arbitrarily to management that they amount to four times the insurable costs is asking for trouble. If management asks for proof, you can only say, "Heinrich said so." Management wants facts—not fantasy. Without proof, hidden costs become fantasy.

In the 5th edition of *Industrial Accident Prevention*, the authors for which are listed as H. W. Heinrich, Dan Petersen and Nestor Roos, these comments are made (81):

> . . . incidental costs of accidents have been estimated to be 4 times as great as the actual costs. There have been innumerable studies made, discussions held, articles written, and arguments presented about that figure. No one has disputed the concept, the fact that there are indirect, or incidental, or hidden costs surrounding accidents. All tend to agree with that. There is tremendous disagreement, however, about how much those costs might amount to. We might first briefly look at what the costs involved are, and then look at some of the controversy over how much they really amount to.

Several of the systems discussed here include property damage costs in their computation of indirect costs. Such property damage costs may or may not be substantial for a particular type of operation. Over time, the impact of property damage costs from hazards-related incidents will relate directly to the investment per employee in the physical plant. Consider these two extreme cases:

- At a public utility, where the cost of investment in the physical plant is $1,000,000 per employee, a turbine rotor was being hoisted and returned to its bearings. A wire cable sling broke. For an injured worker, the workers compensation cost was $1,800; damage to the rotor was $5,500,000.
- In a shop sewing shirts, the investment per employee is about $4,000: high-cost equipment subject to damage doesn't exist, and an indirect cost ratio many times the direct costs for businesses of this sort are not plausible.

In an operation with a very low investment per employee, the possibility of extensive property damage can derive only from what Heinrich called a spectacular event. Further, attention should rightly be given to the construction industry study previously referenced in which the ratio of indirect to direct costs diminishes as the cost of an individual claim increases.

Another point merits discussion, and this is the idea that "additional" sales are needed to cover the costs of injuries. For many years, the author prepared studies on the insurance costs or the direct and indirect costs of worker injuries and illnesses, and concluded with a statement specifying the additional sales necessary to cover them. That has been the common practice for many safety practitioners. Several of the texts referenced in this chapter recommend such presentations. However, the author now doubts the validity of this concept.

As a company goes through a typical budgeting process, both the direct and indirect costs of incidents, knowingly or unknowingly, will be included in estimates of fixed costs, and the prices for products or services would be set to cover those costs. With each dollar of sales, a part of the direct and indirect incident costs are recovered.

Assume that a company projects sales of $10,000,000. Also, assume a profit margin of 3%, and direct and indirect costs at $300,000. To determine the additional sales necessary to cover incident costs, the typical instruction in the literature is to divide the incident costs by the profit margin. Thus, for total incident costs of $300,000 and a 3% profit margin, the additional sales necessary to cover the total costs would be $10,000,000. Computations of this sort appear frequently in the literature. But, their underlying premise is unsound.

No additional sales are needed to cover the expenses for which recovery is obtained through every sale made. The proper relationship for increasing or decreasing incident costs would be as a percentage of earnings at the selected profit margin. For example, a statement to management indicating that incident costs are equal to 100% of profits will receive immediate attention.

DEVELOPING YOUR OWN DIRECT AND INDIRECT COSTS

Obviously, there is a lack of continuity and precision in the commonly used systems on which estimates of direct and indirect costs are based. Nevertheless, reasonably well-researched financial exhibits of direct and indirect incident costs can be impressive and convincing, especially in the prevailing economic climate in which there is so much pressure on managers for financial results.

It is doubtful that a costing system that could stand the test of good science will be available in the foreseeable future. Thus, the author suggests that safety practitioners adapt concepts from the systems discussed here and develop tailor-made approaches to determining worker incident costs in the operations for which they have responsibility.

There are no established formulae that prevent safety practitioners from freely writing their own rules. If an adaptation of the Heinrich, Simonds, or Bird systems is made, a pilot study is recommended to give credibility to the cost system developed and used.

It is not our belief that the previously referenced construction industry study on incident costs and the OSHA program adapted from it will stand the test of good science. But, they do suggest opportunity. They can serve as a base from which safety practitioners can economically produce supportable studies relating directly to their operations.

To begin with, an analysis would be made of a sufficiently broad sampling of the types and numbers of injuries and illnesses reported in the safety practitioner's organization. The average costs of those injuries and illnesses can be produced from actual claims records.

For the non-lost work-day cases, the construction industry study can serve as a reference from which to estimate indirect costs. For the lost work-day cases, the OSHA system could be accessed to produce printouts for the types of injuries and illnesses that have occurred. Such printouts will give direct and indirect cost averages.

The leader of the study should freely make informed cost adjustments as the computations proceed. As an example, if incidents have occurred resulting in

TABLE 6 Estimates of Direct and Indirect Costs of Worker Injuries and Illnesses

Average Case Cost	Number of Cases	Direct Costs	Average Ratio Indirect/Direct Costs	Indirect Costs	Total Costs
		No Lost Work Days			
$375	150	$56,250	5	$281,250	$337,500
		Lost Work-Day Cases			
$1,600	41	65,600	4.5	295,200	360,800
$3,600	9	32,400	1.6	51,840	84,240
$8,000	31	248,000	1.2	297,600	545,600
$18,000	19	342,000	1.1	376,200	718,200
Grand totals		$744,000		1,302,090	$2,046,340

extensive property damage, good judgment should be exercised in determining if and how such costs might be included in the study.

From the accumulated data, plausible and supportable estimates can be made of the direct and indirect worker injury and illness costs specifically related to employer operations. An example of how such a study should be made follows. It is depicted in Table 6.

Assume a 5,000-employee operation, with locations in five states where the medical costs and statutory workers compensation benefits are higher than in most other states. Also assume an OSHA total recordable case rate of 5, and a lost workday case rate of 2. Thus in a year, there would be 150 no lost workday cases, and 100 lost workday cases.

Analysis of the claims cost history for the cases with no lost workdays indicates that most of them fall in the higher cost categories shown in the construction industry study (see Table 3). The average claims cost is estimated to be $375, and one indirect cost ratio, 5, may be used for all of them.

For the lost workday cases, the benefits cost history is on the high side of the ranges shown in Table 5 (taken from the Safety Pays program), but the statistical distribution is a close match. Average claims costs for the four direct cost categories, as computed in the hypothetical study, are $1,600, $3,600, $8,000, and $18,000. Case distribution by cost category is 41, 9, 31, and 19. The ratios of indirect to direct costs given in Table 5 should be used.

With an introduction explaining the methodology used is selecting categories, costs, and ratios, a table such as Table 6 can then be produced. It is an example that could easily be extended to include other data, such as injury types. (As a matter of interest, in this hypothetical example, the ratio of total indirect costs to total direct costs works out to be 1.75 to 1.)

If a safety practitioner undertook such an exercise, a positive response from management would be more likely if a senior-level "champion" bought into the system before the study was commenced.

Part 3
Other Aspects of Financially Related Performance Measures

Much of the literature on reports to management on incident costs relates principally to employee injuries and illnesses. Since the original premises on which the literature is based are seriously out-of-date, some of them dating back to the late 1920s, they do not reflect what has occurred in recent years in terms of the many and complex responsibilities of safety practitioners. It is now the general practice for safety practitioners to have responsibilities beyond those related to safety and health.

Two examples supporting this development are cited here. An annual survey conducted by *Industrial Safety & Hygiene News* and published in December 1996 stated that "about one-third of facilities surveyed operate combined environmental, health, and safety departments at the highest level of the organization." During our interviews with divisional and corporate safety directors, estimates of the number of safety practitioners at locations, other than at divisional or corporate levels, whose jobs included responsibilities for environmental, health, and safety issues varied from 60 to 90%.

These same safety practitioners may have additional consulting responsibilities, such as for auto fleet safety, product liability prevention, public safety, or fire prevention. In any case, all the undesired events in those fields that adversely affect an entity's financial well-being derive from hazards. (Hazards are defined as the potential for harm; hazards include the characteristics of things and the actions or inactions of people.)

Thus, it is the author's suggestion that safety practitioners expand the scope of their reports to management on incident costs to include all hazards-related incidents for which they have a consulting responsibility.

A RADICAL PROPOSAL

Actuarial purists and insurance managers may not approve of the experiment next proposed, but we leave that judgment to the reader. If it was possible to obtain accurate and current worker's compensation, auto liability, product liability, property damage, and other hazards-related claims costs, a trending of that data would be the best performance indicator for most companies, since the data would be expressed in financial terms. Unfortunately, the actual cost of individual claims is not immediately known. For claims reported in a given year, the actual costs may not be finalized for as long as 6 or more years. For those claims, reserves are created for payments that may be made in future years. Those reserves, typically, fluctuate over time. Cost estimates made that include such reserves, understandably, will also fluctuate and their validity is often thus doubted.

If hazards-related incident costs are stated as a fixed number, not subject to later revision because of changes in dollar reserves for particular claims, doubt concerning their reliability will be less frequently expressed. But, how does one obtain a fixed cost number?

Assume that a company's culture has been fairly well established, that its circumstances have remained somewhat stable, that the quality of safety in place has not significantly changed, and that the hours worked represent a fairly sound statistical base. Then, the total hazards-related incident cost figure to be used in our exercise is the total of actual cost outlays in 1 year, *regardless of when the incident was reported.* That would be a radical departure from the typical practice.

The cost figure would include the payments made for incidents reported in previous years as well as in the current year. Any monies held in reserve, regardless of the date the incident occurred, would not be included. Once established, that cost figure would not change for a given year. Adding a soundly based dollar number for indirect costs is appropriate (perhaps using a multiplier of 1.75).

These total costs represent the financial impact of hazards-related incidents on an entity's net operating income for a year. Over the years, the annual cost figures will indicate trends.

CULTURE DEFINED AND ITS SIGNIFICANCE

Why discuss organizational culture in a chapter such as this? Safety practitioners should understand that incident experience and its financial impact are a derivation of the organization's culture, and that unless dramatic action is taken, an entity's culture will change only slowly. This suggests that miraculous and sudden improvements in incident experience are suspect.

What is meant by "culture?" An organization's culture consists of its values, beliefs, legends, rituals, mission, goals, performance measures, and sense of responsibility to its employees, customers, and community, all of which are translated into a system of expected behavior.

Hazards-related incident experience and its dollar implications are a derivation of an organization's culture. How can such a statement be supported? What management does, as a reflection of the system of expected behavior, permeates the thousands of decisions made that create the work environment, set design specifications for facilities and equipment, determine the quality levels to be obtained, establish fire protection standards, respond to environmental needs, etcetera.

When safety is a value within the culture, the senior executive staff displays by what it does that hazards management is a subject to be taken seriously, a subject considered in performance measurement along with other organizational goals. It is at the senior management level that measurable goals are established for performance.

A goal for safety practitioners is to influence an organization's culture as it pertains to safety decision making. Although the method may seem circuitous, one of the purposes of the financial reports they prepare on hazards-related incidents should be to have a favorable impact on the culture.

OSHA STATISTICS AS PREDICTORS OF FUTURE PERFORMANCE

A frequently heard premise is that OSHA recordable incident rates and lost workday case rates are after-the-fact measures and measures of failure; therefore, they are not good indicators of the quality of safety in place and should not be used as indicators of future performance. Those making such statements lack an understanding of the long-term impact of an entity's culture on incident experience, and of the validity of statistical quality-control measurement techniques. If OSHA incident reporting systems are consistent, the trending of rates, upward or downward, should match the trending of injury costs.

The author offers the following statistics to demonstrate the reliability of OSHA rates and as predictors of future experience:

- With 500 employees working 1,000,000 hours a year, the OSHA data are of some but limited value alone: Supplementary performance measures would be needed, such as a workers compensation experience modification, and an independent safety audit.
- With 1,000 employees working 2,000,000 hours a year, the OSHA data begin to have reliability and should be given greater credibility as an indicator of future performance: However, an entity of this size would likely be purchasing workers compensation insurance and should also be considering its experience modification as a performance indicator and arranging for an occasional, independent safety audit.
- At 5,000 employees working 10,000,000 hours a year, the OSHA data have 95.44% reliability as an indicator of future performance.
- Above 5,000 employees, the reliability of the OSHA data improves in small increments, but it never reaches 100%. No predictive system can encompass all of the seldom occurring low-probability, high-consequence events.

Some of those who say that OSHA incident rates are not good indicators of the future suggest that quantifying the number of safety-related activities is a better performance indicator. That process has a great shortcoming since it relies on the assumption that the activities undertaken are directly related to the reality of hazards management needs.

STATISTICAL QUALITY-CONTROL TECHNIQUES

For the safety practitioner in an entity with 1,000 or more employees, the following experiment is proposed. Compute the OSHA recordable and lost workday rates by month, for 24 months. Ask for some assistance from quality-control personnel to plot the numbers on the X Bar within a control chart. Then, secure help in computing upper and lower control limits. Ask whether the data indicate that the system is or is not in control, and whether it is a sound predictor of the future. If the data are valid, continue the computations, adding a month and dropping a month to keep 24 plottings on the chart. When the sample base is large enough, the trends shown should also be an indicator of cost trends.

At 5,000 employees, the data are greatly reliable as a predictor of the future, provided there has been stability in the safety management system. Assume an OSHA recordable incident rate of 5 and a lost workday case rate of 2. For a 1,000-employee entity, the number of OSHA recordables in a year would be 50 and the lost workday cases would be 20. For a 5,000-employee entity, the recordables in a year would be 250 and the lost workday cases would be 100. The data for a 5,000-employee entity have even greater credibility.

In using control charts to determine future experience, it must be understood that if substantive changes in the quality of safety management took place during the term being considered, the data are of limited value. Nevertheless, quality-control charts may be the instrument that influences obtaining the changes needed.

WORKER'S COMPENSATION EXPERIENCE RATING

Interviews with professors in schools awarding safety degrees indicate that many of their graduates are employed in other than large companies. Most of those companies purchase workers compensation insurance, perhaps on a cost-plus basis as provided by retrospective rating plans. For those safety practitioners, the workers compensation experience modification earned by their employer is a significant financial performance measure. The following comments on experience rating first appeared in the author's *On the Practice of Safety* (256):

> The workers compensation experience modification rating system is based on actuarial science. In an undated publication titled "The ABC'S of Experience Rating" issued by the National Council on Compensation Insurance, these statements appear.
>
> Simply stated, experience rating is a procedure utilizing past insurance experience of the individual policyholder to forecast or predict future losses.
>
> In workers compensation experience rating, the actual characteristics of the individual employer are determined over a period of time, usually three years. This experience is then compared with the average as reflected by the manual rate or rates, which apply to the employer's business. If the employer has lower than average costs, then a comparable rate credit is awarded, while for a higher than average experience, a debit rate is applied.

What does all this mean? Actuaries have established that workers compensation claims costs, payrolls, and rates for an insured's occupational classes, over time, form a statistical base from which to compute the future claims experience expected.

Experience rating is one of the historical performance measures than can be used, *cautiously*, as an indicator of the quality of safety in place. Self-insured companies would not have workers compensation experience modifications.

It is not suggested that experience rating is an absolutely accurate indicator of the future. There are many checks and balances in the system, but its preciseness and sensitivity to change may be questioned. And, the system is influenced by one-time events that effect the computations for several years. An example would be the shut-down of an operation that resulted in a rash of claims. Nevertheless, *with caution*, an experience rating modification can be used as one indicator of the quality of safety in place, and of financial cost trending.

CONCLUSION

If safety practitioners are to establish their value, they must be favorably noticed, not only by the people to whom they report, but also by upper management. One means through which they can be favorably noticed, particularly by senior management, is to prepare effective incident experience reports in financially related terms.

Fundamentally, finance is the language of management, and well-crafted, supportable reports with financial implications receive management attention.

This chapter presented a variety of approaches from which safety practitioners can adopt methods customized to their organizations. Patience in obtaining results is a necessity, as well as continuous communication about matters of substance that might achieve the desired results at a given time.

REFERENCES

Accident Prevention Manual for Business and Industry: Administration & Programs, 11th ed. Itasca, IL: National Safety Council, 1997.

Bird, Frank E., Jr. *Management Guide to Loss Control*. Loganville, GA: Institute Press, 1975.

Bird, Frank E., Jr. and Ray J. Davies. *Safety & the Bottom Line*. Loganville, GA: Febco, 1996.

Ferry, Ted S. *Modern Accident Investigation and Analysis, an Executive Guide*. New York: John Wiley & Sons, 1981.

Grimaldi, John V. and Rollin H. Simonds. *Safety Management*, 5th ed. Homewood, IL: Irwin, 1989.

Heinrich, H. W. *Industrial Accident Prevention: A Scientific Approach*, 1st ed. New York: McGraw-Hill, 1931.

Heinrich, H. W., Dan Petersen, and Nestor Roos. *Industrial Accident Prevention*, 5th ed. New York: McGraw-Hill, 1980.

Industrial Safety & Hygiene News, Annual Survey, Dec. 1996.

"Improving Construction Safety Performance." New York: The Business Roundtable, Report A-3, Jan. 1982, reprinted July 1990.

Injury Facts, 1999 ed. Itasca, IL: National Safety Council.

Manuele, Fred A. *On the Practice of Safety*, 2nd ed. New York: John Wiley & Sons, 1997.

OSHA's Safety Pays Program. www.osha.gov, Outreach, OSHA/Software Advisors.

Petersen, Dan. *Techniques of Safety Management*, 3rd ed. Goshen, NY: Aloray, 1989.

3

The Theoretical Ideal for Optimum Safety Performance

INTRODUCTION

In time, an occupational safety management standard will be issued by some organization. Discussions continue throughout the world on developing such a standard. In 1996, the International Organization for Standardization held a meeting in Geneva at which the idea was considered but no concensus was reached. At the time, there was talk of having the standard follow the basics of the ISO 9000 series on quality management. A proposal for an ISO safety standard will likely appear again.

More recently, the British and the Australians have advanced their own ideas on what a proper safety management standard should be. OSHA has also made several starts toward issuing a standard, and that activity continues. In December 1999, the Industrial Hygiene Association received final approval from the American National Standards Institute of its application for accreditation to serve as the secretariat to develop a standard pertaining to occupational safety and health management.

The foregoing comments pertain only to recent attempts to write the definitive paper on the elements of safety management. Activity of this sort has been on-going for many years. For example, in 1973, the American Society of Safety Engineers (ASSE) proposed to the National Institute for Occupational Safety and Health that it be engaged to develop "a statement defining the basic elements and sub-elements of an effective Occupational Safety Program" (3).

In that proposal, ASSE acknowledged a situation that still exists, one which will have to be faced by any organization attempting to draft a standard:

> Considerable confusion currently exists because of the wide range of viewpoints on the basic requirements of a total safety and health program. Numerous establishments are currently attempting to adopt or establish safety and health programs, but are hampered by the multiplicity of viewpoints (5).

When a safety management standard is written, those applying it should expect to achieve optimum safety performance. However, they will not be confident in the standard if it is anything like typical safety management outlines. An example of such an outline is the *Safety and Health Program Management Guidelines* issued by OSHA in 1989, which are, in effect, the criteria for acceptance in the agency's Voluntary Protection Program.

In crafting a safety management standard for this century, it is hoped that those involved will:

- Consider the theoretical ideal for optimum safety performance
- Develop an awareness of the inadequacies that characterize less than stellar performance
- Have the capability of discarding that which is no longer applicable
- Make revisions and additions to the typical safety management guidelines, reflecting what has been learned

Figure 1 presents a model for the theoretical ideal in Safety Performance. *Only the major elements* are illustrated, with supporting rationale.

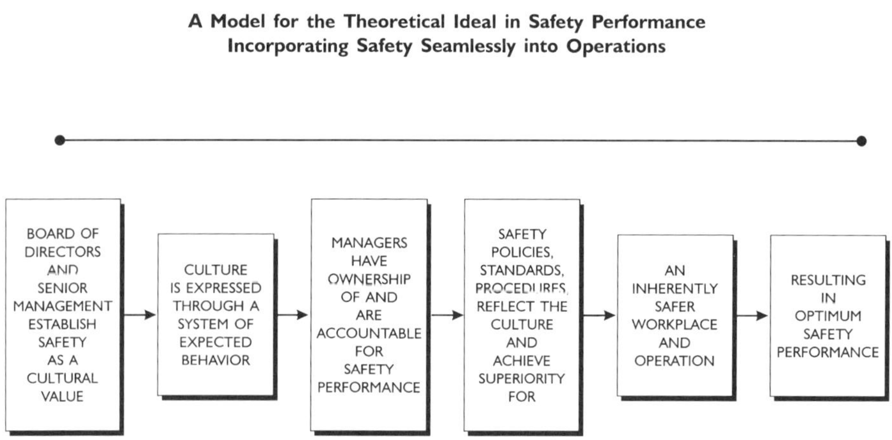

FIGURE 1 *Model for the theoretical ideal in safety performance.*

A PERSONAL AND HISTORICAL VIEW

Over many years, the author has assisted a number of clients in writing draft statements on safety policy, the procedures for implementing them, and safety program management guides. Hindsight suggests that most of these efforts were lacking in substance and effectiveness in some respects. Greater emphasis should have been placed on the relationship between an organization's demonstrated culture and safety results, the accountability system, and the procedures necessary to achieve an inherently safer workplace.

CULTURE: ITS SIGNIFICANCE

Safety is culture driven. That is now a given. It is a certainty that superior performance is achieved only when safety is truly valued within an organization's culture. Safety policy statements, no matter how persuasively written, come up short on effectiveness if they are not supported by a culture that has made safety a core value.

It has been well documented that the safety results achieved, favorable or unfavorable, have a direct relationship to the status of safety within an organization. Judith Erickson's article entitled, "Corporate Culture: The Key to Safety Performance" is one example of a pervasive theme in the recent literature: All that gets done or does not get done in an organization is a derivation of its culture. As Erickson said:

> It is important for safety professionals to understand that a company's culture forms the basis for everything that is valued in the company, including safety (45).

Erickson's statement is in the same class with comments others have made in recent years, and it represents the hard fact of reality. For example, consider the opening sentences in an article entitled, "Culture Is the Foundation for Safe Behavior" by Jerry L. Burk. They give his view on the significance of culture:

> It is imperative that culture be the first consideration when implementing safe behavior practices within an organization. Changes in cultural dimensions can influence behavior: however, changes in behavior do not have the same impact on culture (84).

Literature on the subject is abundant, but establishing safety as a core value within an organization's culture is not the norm. Policy statements may say that safety is important, but experience has shown that such statements may be superficial and have little real effect on what gets done if safety does not have core value status.

To achieve world-class results, safety must be established as one of the most important values within a culture. This is made clear in the National Safety Council's 450-page book entitled, *Safety Culture and Effective Safety Management*, published in July 1999.

CULTURE DEFINED

In an article entitled, "Behavioral Safety and Future Trends for Change Management," Jerry Pounds offered this ever so brief description of an organization's culture:

> A culture is, at its core, the sum total of those behaviors that leadership rewards and punishes.

Although this definition uses few words in comparison with other definitions of culture, it speaks directly to one of the themes in this essay: If management's behavior concerning safety is not one of the criteria for which leadership rewards or punishes, safety is not a core value.

A more detailed definition of culture, a composite of the viewpoint expressed by many writers, follows:

> An organization's culture consists of its values, beliefs, legends, rituals, mission, goals, and performance measures, and its sense of responsibility to its employees, its customers, and its community, all of which are translated into a system of expected behavior.

Erickson made these related comments on values, and how a culture is maintained:

> These values are manifested in management actions and behavior. They are reflected by what management does, pays attention to, ignores, measures and controls. It is through management actions and behavior that employees become aware of the organization's meanings and learn what is expected of them and how to behave. Through a reward and punishment system, the culture in an organization is maintained (45).

CULTURE TRANSLATES INTO A SYSTEM OF EXPECTED BEHAVIOR

In the definition of a culture just previously given, the last phrase resonates: The elements of the culture are translated into a system of expected behavior, and the behavior pertaining to safety will be what the system expects, rewards, and punishes. That behavior defines the level of management commitment to safety, and whether safety has achieved the status of a value.

At all levels, the staff determines what it perceives to be important to their superiors and acts accordingly. This is a significant aspect of superior safety management. A system of expected behavior evolves among the management staff and permeates the entire organization. All employees begin to understand what is expected of them in terms of safety considerations. They will not be easily fooled by hollow rhetoric or meaningless actions on the part of management.

Management is defined by its actions. Management is what management does, which may be different from what it has published or endorsed as a safety policy, procedure, or program. What management does or does not do expresses its system of expected behavior and represents the reality of its safety culture. If management wants superior safety results, it must walk the talk. Safety must truly be one of the entity's core values, and this must be demonstrated by its system of expected behavior.

MANAGEMENT OWNERSHIP AND ACCOUNTABILITY FOR SAFETY PERFORMANCE

Managers must have ownership of and be held accountable for safety performance. Procedures written for the implementation of safety policies are deficient if they do not establish accountability for hazards-related incident experience as a factor in the management performance evaluation system. That is, performance accountability ultimately has an impact on job retention, promotability, and total remuneration.

Management holds people accountable for what is important in its system of expected behavior and measures them accordingly. As the author wrote in *On the Practice of Safety*:

> There are at least two classes of measures—measures of form and measures of substance. Measures of form are just that. Measures may be produced and observed, such as safety audits, incident statistics, or incident costs, and have no bearing on the status of those responsible for operations. Measures of substance have an impact on the well being of those responsible for results as a part of an overall accountability system (221).

A simple but true test of whether safety is an established value within a culture is whether accountability for safety results is, in reality, a performance measure. If this is not the case (and the author is well aware that most often it is not), an organization does not value safety equally with other performance goals, no matter what may be specifically stated in its safety policy.

Early on in most safety program management guidelines, it is clearly established that management commitment is vital. Good examples are the following OSHA issuances:

- *Safety and Health Program Management Guidelines* issued in 1989 (see Appendix A)
- The several proposed safety management standards
- *Ergonomics Program Management Guidelines For Meatpacking Plants* issued in 1999
- General industry ergonomic standard issued in 2000

Although management's commitment to safety is important, it is suspect if the accountability system does not include safety performance measures that affect the well-being of those responsible for results.

DEVELOPING SUPERIOR POLICIES, STANDARDS, AND PROCEDURES

The quality of the safety policies, standards, and procedures in place is a reflection of the organization's culture. So, also, is the communication about them. In companies that have attained outstanding safety performance, it is common for the standards and procedures applied to exceed the requirements of published standards, such as those of ANSI or OSHA.

When the culture drives safety, a positive discipline emerges, reflecting the system of expected behavior, in which individuals in the workforce follow certain procedures and standards, and expect others to do the same. The development and application of more comprehensive policies, standards, and procedures result in an inherently safer workplace and operations.

AN INHERENTLY SAFER WORKPLACE

The data are persuasive: The greatest strides forward in reducing safety, health, and environmental risks are being made in the design processes. A safety management guide is deficient if it does not contain design and equipment-acquisition procedures that must be followed to avoid bringing hazards into the workplace, which results in an inherently safer workplace.

Chapter 11, "The Safety Through Design Movement," describes the benefits obtained by considering hazards and risks early in the design process. Chapter 7, "How to Avoid Bringing Hazards into the Workplace," offers suggestions on how to do what the title implies—avoid bringing hazards into the workplace. Appendix A at the end of that chapter is an adaptation of Intel's philosophy and the methods used to "Design for Environmental, Health and Safety."

We have learned that retrofitting to eliminate or control hazards inadvertently brought into the workplace is very expensive, and sometimes less than adequate. Also, we have learned that, for business reasons, retrofitting may not be considered practical. When hazards are not to be eliminated or mitigated by retrofitting, controlling the risks resulting from such hazards is only partially successful when all that can be applied are administrative and personnel management procedures. We can, and must, do better than that.

Several examples given in this book illustrate how companies have established relationships with design engineering firms and equipment suppliers that result in the design of equipment and purchase of products with very high safety standards. The benefits of this sort of relationship are significant.

Erickson has made much of the tangible expressions of corporate culture:

> Organizational factors, such as tangible expressions of corporate culture, are the primary determinant of the level of safety performance in an organization (46).

The most tangible evidence of safety's importance within an organization's culture is the level of safety built into its facilities, hardware, equipment, tooling, machinery, operations, and processes. Appearances reveal a great deal about how successful an entity avoids bringing hazards in the workplace and whether it really intends to create an inherently safer workplace.

Where a culture is sound with respect to safety, it will be obvious that management intends to provide "employment and a place of employment which are free from recognized hazards"—to borrow an OSHA phrase. In light of what we have learned in recent years, it makes no sense in the twenty-first century for a safety management plan not to include safety through design procedures: i.e., methods to avoid bringing hazards into the workplace and to attain an inherently safer workplace.

SAFETY IN OPERATIONS

Superior policies, standards, and procedures are the foundation on which safety in operations is built. It must be understood that an inherently safer workplace and outstanding operational safety are complementary: They are necessary parts of a whole.

Ownership of safety and keeping risks at an acceptable level in operations are specifically a reflection of an organization's system of expected behavior.

OPTIMUM SAFETY PERFORMANCE

Although we support the premise that all injuries are preventable, the goal of an ideal safety performance model is optimum results, not perfection. The probability that an organization can go a year without a recordable injury is inverse to the number of its employees.

Unfortunately, a single measure does not exist that can determine when outstanding results are achieved because the risks of each type of operation differ. However, using the principles of quality assurance as our basis, we can attempt to determine whether optimum performance, or near-optimum performance, has been achieved. *Six Sigma* has become a popular standard in quality-assurance circles. According to Six Sigma, quality performance exists when no more than 3.4 defects per million occur. Assume that an OSHA recordable injury is the defect. OSHA recordable rates are predicated on the basis of 200,000 hours worked: 200,000 is one-fifth of a million. An OSHA recordable rate of $0.68 \times 5 = 3.4$. Thus, an OSHA recordable rate of 0.68 is synonymous with a Six Sigma performance level.

Meeting Six Sigma requirements is achieved in companies where operations are considered inherently risky and safety is a core value within their culture.

MOVING FORWARD

How can an organization establish safety as a core value within its culture? Not easily and not quickly. However, a new force has emerged that may convince more managements to examine the status of safety in their organizations. Safety practitioners, take heed.

In April, 2000, the National Safety Council published its *Safety Agenda For The Nation*. The Agenda "is designed to call the nation's attention to the seriousness of injury problems, and to propose effective actions that target the largest causes of unintentional injury deaths." One section addresses safety and health in the workplace. The objectives the Council outlines for senior management are precisely in line with what the author believes should be the goals of safety practitioners as they move into the twenty-first century. The following are excerpts from "The Solution" proposed in the Council's report:

> Implementing a systematic approach to workplace safety will require a cultural change in many organizations, among regulators and within the safety profession. All organizations need to nurture a "safety culture." The Council will ask senior management of public and private companies to institute and communicate to all employees a *Corporate Code of Safety and Health Ethics*. The code will:
>
> - Add safety and health to the **core values of the organization**, on the same level and with the same support as other values defined by the organization, such as customer service, financial performance and productivity.
> - Recognize that the Chief Executive Officer (CEO) or public equivalent is the leader of the organization, and all of his or her actions and decisions add to or detract from the **credibility of the safety and health commitment**.
> - Ensure that the CEO and senior management **personally set the standard** for safety and health performance.
> - Establish a **comprehensive evaluation**, a safety and health audit of the organization to identify existing and evolving hazards and to develop an effective system of accountability to ensure they are controlled or abated.
> - Recommend that the corporation's board meeting and annual report to shareholders or other stakeholders include a discussion of the company's **safety and health performance**.

These are powerful suggestions. If management does as the Council proposes, deficiencies in safety policies and the procedures to implement them would be largely overcome.

Safety practitioners may be asked to determine how their organization functions in relation to the Council's agenda, or they may want to use the agenda as a base from which to initiate a comparative study of their own. In either case, the enormity of what is being proposed should be understood.

CHANGING A CULTURE

Estimates in the literature of how long it generally takes to change an organization's culture vary from 4 to 8 years. Doing so for safety is not easy to accomplish. Success requires, first, a decision by management that safety is a core organizational value, and a permeation of that idea throughout the senior and middle management staff.

Such a decision then must be followed by continuous operational evidence over the long term to achieve a belief at all levels that safety has become a core value. The beliefs of the staff will only change as a result of management personal actions; a fundamental revision of some aspect of the performance accountability system that becomes well known; and the successful completion of certain needed actions as defined by a culture assessment.

ASSESSING THE CULTURE

If a change in culture is to take place, one that results in safety becoming a core value on the same level and with the same support as other values, proven problem-solving techniques should be applied. The first step is to define the problem. That requires making an assessment to determine any shortcomings in safety management.

Initial assessments can be made internally, for which the safety practitioner serves as a consultant. Or, external resources can be brought in to lead the entire project. Making an initial evaluation has the advantage of alerting management at all levels to real problems and to what may be found in a perception survey conducted by a consultant-specialist in that field.

Very often, there is a marked difference between what management thinks is the reality of safety management and the perceptions of employees. Fortunately, it is not difficult to determine the perceptions employees have about the importance safety has in the management of a business. To obtain an initial indication of what employee perceptions might be and to define the cultural problems to be overcome, it is suggested that an internal evaluation be made of seven basic elements of safety management.

Safety practitioners who take the initiative in such an endeavor will need to obtain management's approval of an appropriate, cross-functional team to conduct such a review. The goal would be to establish the quality of what actually takes place, which is the reality base on which people establish the perceptions and beliefs they have about safety.

The findings of the study are the outcomes of, and define the organization's system of expected behavior. But to employees, these outcomes are causal factors that determine their beliefs. The review team should take the following self-analytical steps:

1. Assess the procedures in place through which employees are empowered to eliminate or mitigate physical and operational hazards, or refer them to management for attention. Determine how well the procedures work: Do employees use the system and have faith in it; what actions are taken, and how soon; and how are any conclusions communicated?
2. Review the inspection program to determine whether it is thorough and effective, whether follow-up takes place, and when. Consider the following: Is the inspection program substantive or superficial; are known hazards addressed or overlooked; do the same hazards appear repeatedly; are findings resolved soon or over considerable time; and are communications on the results adequate or less than adequate?
3. Review the incident investigation system and rate it for quality (see Appendix B, "Quality of Incident Investigation Self-Audit Outline," provided at the end of this chapter). Above all, determine whether there is a balanced emphasis in the investigation system on employee actions, and on physical and operational hazards beyond the control of employees.
4. Review the process by which safety-related work orders are sent to maintenance or engineering: Determine whether the priority given to them is adequate or less than adequate.
5. Evaluate the effectiveness of safety committees. Ask whether the committee has a charter, it receives training, and considers real problems or meets as a matter of form with little effectiveness; members consider serving on a committee worthwhile; and what happens to the solutions the committee proposes.
6. Assess the safety training given (worker, supervision, management): Explore whether training addresses real needs, and whether any needed retraining is scheduled and then given.
7. Evaluate the measures taken to identify and resolve ergonomics problems, including employee comments about overly stressful work. (This technical subject is included in our list because of the prominence of musculosketal injuries among those reported nearly everywhere.)

Why choose to assess these seven elements in safety management? How well they are managed is much of what employees notice, and what employees see forms their perceptions of the status that safety has within the organization. These fundamentals must be done well to establish in the minds of employees that safety is a value.

In an entity where safety results have not been stellar, a safety practitioner advising the team to do a thorough job in making this initial evaluation should not be surprised if their reporting shows real shortcomings in these basic management systems. At this point, regardless of the results of the survey, a report to management on the team's findings should be prepared, after a thorough verbal briefing.

POSSIBLE ACTIONS AFTER AN INITIAL EVALUATION

Management may decide that the findings of the team conducting the initial culture assessment are a sufficient basis upon which to build a strategy. If the problem identification activity is to be extended, at least four possibilities exist:

- A thorough, internal culture-determining safety audit may be conducted.
- An outside consultant may be engaged to conduct a complete safety audit.
- A perception survey may be conducted internally to give employees an opportunity to record their beliefs.
- An outside consultant may be brought in to do a complete perception survey.

Before a complete safety audit or perception survey is made, management must have decided that the results of the audit or the employee perceptions will be addressed and that the entire staff will receive some sort of communication representing management's response. Conducting a perception survey without good feedback will more than likely produce negative results, and reinforce any existing beliefs about safety not being important to management.

As part of a good problem-solving technique, a follow-up decision should be made on which solutions to apply. An effectiveness evaluation should also be made after an appropriate period of time.

SAFETY AUDITS: AN INSTRUMENT OF CHANGE

If a safety audit system is to achieve what Kase and Weise suggest is the mark of success of any auditing program, it must serve as the basis for decision making on the part of management pertaining to the organization's culture. As Kase and Weise state in *Safety Auditing: A Management Tool*:

> Success of a safety auditing program can only be measured in terms of the change it effects on the overall culture of the operation, and enterprise that it audits (36).

The literature on safety audits is plentiful, as are the definitions and purposes of safety audits. It is the author's belief that the paramount purpose of a safety audit is to provide an assessment of the quality of safety in place, which is a reflection of the expected system of behavior. Thus, the audit report must contain assessments of the effectiveness of the policies, standards, procedures, and systems that should be in place.

Hazards noted, physical or operational, are to be viewed as being representative of possible deficiencies in management systems. A management that wants to achieve a change in culture is best served if the deficiencies noted and the proposals made in audit reports for improvement principally effect safety management systems.

The safety audit report that merely lists physical and operational hazards to be corrected is of little value if the real goal is to improve the safety culture. Even if every proposal made for specific hazard elimination or control was given appropriate attention, little would be gained in the long run if the changes in the systems of expected behavior that permitted the existence of the hazards were not made.

In those companies that have internal staffs which make safety audits, it is suggested that an examination of their purposes be made to ensure that they address safety culture needs. It might be found that the audit guide instruction and protocol prompt more of an inspection than an assessment of how effective the safety management systems are.

In addition, if outside consultants are engaged to make safety audits, the interests of the company responsible for the hiring would be best served if a clear understanding was reached on what the audit is to accomplish in identifying system and culture change needs. The consultant's audit guide outline should also be reviewed for relevancy. To paraphrase Kase and Weise, the purpose of the safety audit is to effect the overall culture of an organization.

CULTURE PERCEPTION SURVEYS: THEIR SUBSTANCE

Executives in some companies with which the author is familiar have engaged consultants, with good results, to assist in achieving necessary culture changes. Those attaining the greatest success addressed all levels of employment, but also gave primary emphasis to what management needed to do to overcome a well-founded negative image concerning safety.

Some consultants will provide assistance in crafting a perception questionnaire to be presented to employees entirely under the direction of the internal staff, their purpose being to allow employees to confidentially express their views on safety. Nonetheless, there are real advantages to having the perception survey led by outside consultants who are particularly skilled in the field: One is that the surveyors can place an emphasis on top management more effectively than the internal staff; the other is the additional credibility given to the results of their assessment.

Consider the following example: A consultant receives a call from a safety director requesting a safety audit; the safety director then proceeds to describe the shortcomings in safety management at his company. When asked why he wanted to spend the money on an outside consultant to tell him what he already knew, his response was this: The report will be on the consulting firm's stationery and it will more likely get the attention of management.

Just what are perception surveys and what purpose do they serve? In a chapter entitled, "Achieving the Necessary Culture Change" in *Safety Through Design* (37), Steven Simon makes the following comments:

> There are six phases in the culture change model. They apply to any organization-wide culture change. The six steps are as follows:
>
> - Define and communicate the need for change;
> - Envision a desired future;
> - Assess the culture;
> - Strategic planning;
> - Implement; and
> - Evaluate.

According to Simon, the first step in a culture change model is analysis and then communication on the need for change. Subsequently providing a vision helps people let go of the past: They need to believe that something better will replace what they have now. To assure that an organization addresses problems and not symptoms, Simon proposes a time-saving assessment of culture. The remaining three steps—strategic planning, implementation, and evaluating results—fulfill the requirements of a sound problem-solving technique.

Simon closes his discussion with these words of caution that apply to all levels of management:

> In order to achieve the necessary safety culture change, we not only have to apply our technical expertise, we must also identify and remove negative safety norms, lead by personal example, and engage and inspire others to share a vision of a new culture. Although this may be a difficult process, it is the best and only guarantee of sustainable success.

Simon is also responsible for developing the *Simon Open System (S.O.S.) culture change model* (194):

> It is designed to evaluate the safety process by dealing with a wide range of leading indicators to safety performance, inclusive of both the structural and technological factors that go into creating a safe environment, and the cultural influences that shape safety norms or behaviors. It develops a profile of behaviors and supports within the organization that affect its ability to manage safety efforts in order to provide a road map to design strategy for performance improvement.

Simon uses the *S.O.S. Culture Perception Survey* as an instrument in support of his culture change model. Through its use, quantitative measurements are provided of such "soft" issues as a belief in management's commitment to safety, trust, and communication.

Another perception survey, developed by Erickson Associates, is called the *Holographic Safety Net*. According to the company literature,

> The Erickson Associates perception survey investigates the effect of organizational factors on safety performance. These organizational factors include: organizational structure; organizational importance of safety and health; safety responsibility and accountability; and employee responses, behavior, and perceptions.

Perception surveys provide management with a clear picture of what employees at various levels of responsibility think about the effectiveness of their management of safety, and propose solutions that are directly relevant to its operations.

Consultant Dan Petersen has stressed, among other factors, the significance of accountability for safety being well established—at all levels. As the principle presenter in the "Safety Accountability" program, he regularly stresses the following message to senior management:

> To gain control over your safety processes and outcomes, an organization must manage safety in the same way it manages production and quality—by making each individual accountable.

Literature on the program instructs senior management that the overall culture of an organization should be measured. This is Petersen's view of perception surveys:

> After many years of development and testing, we found that a perception survey of employees and managers in an organization gave us a better prediction of the future of the safety record than any other indicator we tested. We have been using perception surveys for a number of years and found their use to be invaluable in diagnosing what needs to be done to improve safety systems in organizations.

As a part of the overall "Safety Accountability" program, a separate manual is provided, the purpose of which is to "describe how to use CoreMedia's Dan Petersen Safety Perception (SPS) software program package." Its purpose is to identify shortcomings in safety management systems and to provide information to management on incorporating improved methods in the accountability system.

CONCLUSION

As stated here, achieving a change in an organization's culture is not easy to do. A colleague once commented that the time required to achieve a culture change is inversely proportional to the age of the business. Nevertheless, the facts are clear: If superior results in safety are to be attained, it must be a core value within an entity's culture.

Of course, the goal of all our research and analysis is to arrive at the theoretical ideal for optimum safety performance. This is achieved when, because of the influence of an organization's culture, safety is seamlessly woven into all decision making, all risks deriving from hazards are at an acceptable level, and all employees perform their jobs safely. Is the theoretical ideal ever reached? No, but it is still a laudable goal. Getting close to it will occur only when an entity's culture establishes safety as a core value.

This chapter commenced with the statement that, in time, an occupational safety management standard will be issued. Quite probably, such a structure will resemble the *Safety and Health Program Management Guidelines* issued by OSHA in 1989. Excerpts from those Guidelines, in Appendix A here, indicate their principal elements and recommended actions. However, it is our hope that the concerns of safety professionals will prevail and the standard eventually developed will incorporate much of the theoretical ideal.

REFERENCES

Burk, Jerry L. "Culture Is the Foundation for Safe Behavior." *Safety + Health*, May 2000.

Christensen, Wayne C. and Fred A. Manuele. *Safety Through Design*. Itasca, IL: National Safety Council, 1999.

"Develop a Definition of Basic Elements, Sub-Elements, and Guidelines for an Effective Occupational Safety Program." Unpublished contract proposal to the National Institute for Occupational Safety and Health. Des Plaines, IL: American Society of Safety Engineers, May 1973.

Ergonomic Program Management Guidelines for Meatpacking Plants. Washington, DC: U.S. Department of Labor, 1990, OSHA 3123.

Erickson, Judith A. "Corporate Culture: The Key to Safety Performance." *Occupational Hazards*, April 2000.

Kase, Donald W. and Kay J. Wiese. *Safety Auditing: A Management Tool.* New York: Van Nostrand Reinhold, 1990.

Manuele, Fred A. *On the Practice of Safety.* New York: John Wiley & Sons, 1997.

Perception Survey. Irvine, CA: Erickson Associates, 1999.

Petersen, Dan. *Safety Accountability*. Portland, OR: CoreMedia, 1999.

Pounds, Jerry. "Behavioral Safety and Future Trends for Change Management." *Performance Management Magazine*, Vol. 16, No. 1.

Safety Agenda for the Nation. Itasca, IL: National Safety Council, 2000.

Safety and Health Program Management Guidelines. OSHA, Federal Register, 54 FR 3904, Jan. 26, 1989.

Simon, Steven I. "Achieving the Necessary Culture Change." *Safety Through Design.* Itasca, IL: National Safety Council, 1999, Chap. 4.

Simon, Steven I. *Proceedings: A Behavioral Safety Symposium.* Des Plaines, IL: American Society of Safety Engineers, 1998.

Swartz, George, ed. *Safety Culture and Effective Safety Management*. Itasca, IL: National Safety Council, 1999.

Appendix A

OSHA's Safety and Health Program Management Guidelines

SCOPE AND APPLICATION

This guideline applies to all places of employment covered by OSHA, except those covered by 29 CFR 1926, which is the construction safety standard.

INTRODUCTION

Effective management of worker safety and health protection is a decisive factor in reducing the extent and severity of work-related injuries and illnesses.

Effective management addresses all hazards whether or not they are regulated by government standards.

OSHA urges all employers to establish and to maintain programs which meet these guidelines in a manner which addresses the specific operations and conditions of their worksites.

THE GUIDELINES—GENERAL

Employers are advised and encouraged to institute and maintain in their establishments a program which provides systematic policies, procedures, and practices that are adequate to recognize and protect their employees from occupational safety and health hazards.

An effective program includes procedures for the systematic identification,

Source: Excerpts from Federal Register, 54 FR 3904, Jan. 26, 1989.

evaluation, and prevention or control of general workplace hazards, specific job hazards, and potential hazards which may arise from foreseeable conditions.

An effective program looks beyond specific requirements of law to address all hazards.

Written guidance ensure(s) clear communication of policies and priorities and consistent and fair application or rules.

[Next in the Guidelines comes item (b), Major Elements, of which there are four. In a following item (c), Recommended Actions are set forth for each of the Major Elements. In these excerpts, the applicable Recommended Actions follow Major Elements, directly.]

MAJOR ELEMENT 1

Management commitment and *employee involvement* are complementary. Management commitment provides the motivating force and resources and applies its commitment to safety and health protection with as much vigor as to other organizational purposes.

Employee involvement provides the means through which workers develop and/or express their own commitment to safety and health protection, for themselves and for their fellow workers.

Recommended Actions

State clearly a worksite policy on safe and healthful work and working conditions, so that all personnel with responsibility:

Understand the priority of safety and health protection in relation to other organizational values.

Establish and communicate a clear goal for the safety and health program and objectives for meeting that goal.

Provide visible top management involvement.

Provide for and encourage employee involvement.

Assign and communicate responsibility so that [all] know what performance is expected of them.

Provide adequate authority and resources.

Hold managers, supervisors and employees accountable.

Review program operations at least annually to evaluate their success so that deficiencies can be identified. [and] objectives can be revised.

MAJOR ELEMENT 2

Worksite analysis—involves a variety of worksite examinations, to identify not only existing hazards but also conditions and operations in which changes might occur to create hazards.

Effective management actively analyzes the work and worksite, to anticipate and prevent harmful occurrences.

Recommended Actions

Conduct comprehensive baseline worksite surveys for safety and health and periodic comprehensive update surveys.

Analyze planned and new facilities, processes, materials, and equipment.

Perform routine job hazard analyses.

Provide for regular site safety and health inspections.

Provide a reliable system for employees to notify management personnel [of perceived hazards] without fear of reprisal.

Provide for investigation of accidents.

Analyze injury and illness trends.

MAJOR ELEMENT 3

Hazard prevention and control—where feasible, hazards are prevented by effective design of the job site or job. Where it is not feasible to eliminate them, they are controlled to prevent unsafe and unhealthful exposure.

Recommended Actions

So that all current and potential hazards, however detected, are corrected or controlled in a timely manner, establish procedures for that purpose, using the following measures.

Engineering techniques where feasible and appropriate.

Procedures for safe work which are understood and followed—as a result of training, positive reinforcement, and, if necessary, enforcement through a clearly communicated disciplinary system.

Provision of personal protective equipment.

Administrative controls, such as reducing the duration of exposure.

Provide for facility and equipment maintenance so that hazardous breakdown is prevented.

Plan and prepare for emergencies, and conduct training and drills.

Establish a medical group.

MAJOR ELEMENT 4

Safety and health training—addresses the safety and health responsibilities of all personnel concerned.

Recommended Actions

Ensure that all employees understand the hazards to which they may be exposed and how to prevent harm to themselves and others from exposure to these hazards.

So that supervisors will carry out their safety and health responsibilities effectively, ensure that they understand those responsibilities and the reasons for them, including:

- Analyzing the work to identify hazards.
- Maintaining physical protection in their work areas.
- Reinforcing employee training on hazards [and] on needed protective measures, through continual performance feedback and, if necessary, through enforcement of safe work practices.

Ensure that managers understand their safety and health responsibilities as described under Management Commitment and Employee Involvement, so that the managers will effectively carry out those responsibilities.

Appendix B

Quality of Incident Investigation Self-Audit Outline

1. First, what incident causation model is to apply must be determined. All that follows in this outline regarding causal factors will relate to that model.
2. Taking a sample of incident investigation reports completed by supervisors, rate the following for quality.
   a. General information data
   b. Incident descriptions
   c. Recording of injury and illness data
   d. Completion of code entry requirements
   e. Causal factors determinations
   f. Actions to be taken to prevent recurrence
   g. Completion of questions particular to the organization
   h. Generally, the quality of the report, overall
3. Using the same sampling of reports, make these assessments.
   a. Are the terms on the forms understood?
   b. Are requirements for causal entries on the forms too simplistic?
   c. Do requirements for causal entries fit with the concepts of incident causation to be applied?
   d. Does the quality of the reports show a need for additional training in incident investigation?
   e. Do the contents of forms indicate a need for additional job hazard analyses?

Source: Excerpts from Fred A. Manuele, *On the Practice of Safety*, New York: John Wiley & Sons, Chap. 9, "Designer Incident Investigation."

   f. Does the data gathered allow meaningful analyses?
   g. Is the needed managerial attention given to reports?
   h. Are safety personnel fulfilling their responsibilities?
4. Concerning written incident investigation procedures or guidelines, and the incident investigation report.
   a. Do they clearly convey an understanding of the causation model to be applied?
   b. Are the definitions clear? Is the language ambiguous?
   c. Do they properly balance consideration of design and engineering practices, and management, operations, and personnel task performance practices?
   d. Do they clearly establish which incidents are to be investigated?
      - Personal injuries (at what level)
      - Property damage (at what level)
      - Environmental incidents (at what level)
      - A frequency of minor injuries, property damage, or environmental incidents
      - Incidents not resulting in injury or damage, but which may have in other circumstances
   e. Are responsibilities for investigation clearly defined?
      - Supervisors
      - Upper management
      - When incident teams are to be used
      - Safety personnel
   f. For report distribution
      - Are the approval levels what they should be?
   g. For corrective action
      - Are responsibilities precisely established?
      - Is the follow-up procedure appropriate?
   h. Do the procedures or guidelines and the incident investigation form need revision and re-issuance?
5. Concerning training in incident investigation.
   a. Should the program be improved?
   b. Has it been given to all who need training?
   c. Are refresher courses given?
   d. Is an additional focus on training needed?
6. Is there a need to convince management that giving greater significance to quality incident investigation will assist in their achieving their goals?
7. Having made this self-audit, what actions are to be taken to improve the system?

4

The Safety Decision Hierarchy

INTRODUCTION

A hierarchy is any system of persons or things ranked one above the other. Within recent years, exhibits labeled a "hierarchy of controls" or a "safety hierarchy" have appeared in several safety publications. These hierarchies identify the actions to be considered, in an order of effectiveness, to resolve hazard and risk situations. They represent an important step in the evolution of the practice of safety, even though they lack the opening and concluding actions that are characteristic of sound problem-solving and decision-making techniques.

A typical hierarchy of controls would list the elements shown in Table 1, in a descending order of effectiveness. To serve the cause of good safety management, hierarchies of control need to be extended to include three other elements found in good problem solving and decision making techniques. These elements are: problem identification and analysis; measurement of results for effectiveness; and reanalysis when results are inadequate. The *Safety Decision Hierarchy* presented here adds substance to the typical hierarchy of controls and extends the concept to meet the needs of good problem solving.

TABLE 1 A Typical Hierarchy of Controls

Eliminate the hazard or risk.
Apply engineering controls.
Provide warnings.
Train and instruct.
Provide personal protective equipment.

A REVIEW OF THE LITERATURE

Problem-solving approaches have become somewhat standard, and the related methodologies noted in a number of texts are similar. As Kepner and Tregoe state:

> ... The elements of problem solving and decision-making do not change. Only the situations change (Introduction, x).

An example of the problem-solving techniques typically shown in the literature follows in Table 2. The model was chosen particularly for its simplicity, but it is nevertheless complete. Note especially that problems must be clearly identified and defined, and the effects of the actions taken are to be carefully scrutinized.

TABLE 2 The Ideal Approach to Problem Solving

I =	identifying problems
D =	defining problems
E =	exploring alternative approaches
A =	action on a plan
L =	looking at the effects

Source: John D. Bransford and Barry S. Stein, *The Ideal Problem Solver*, New York: W. H. Freeman and Company, 1984, p. 11.

A broad review of safety literature was made to determine whether all the steps that need to be taken to address and resolve safety issues were set forth by an author in a logical, problem-solving progression. One model that initially seemed to meet these criteria was that developed by William G. Johnson in *MORT Safety Assurance Systems* (135). Influenced by the principles Kepner and Tregoe outlined in *The Rational Manager*, Johnson presented a *performance cycle*:

> The cycle, a distillation from a repertory of management methods useful in safety, mirrors the safety system which enhances safety's value to management (139).

We illustrate Johnson's cycle in Figure 1, with the addition of one crucial action: a consideration of alternatives. As in any effective problem-solving procedure, the process commences with problem identification and analysis.

THE SAFETY DECISION HIERARCHY

The safety decision hierarchy in Figure 2 presents a logical sequence of actions encompassing the problem-solving and decision-making techniques that safety

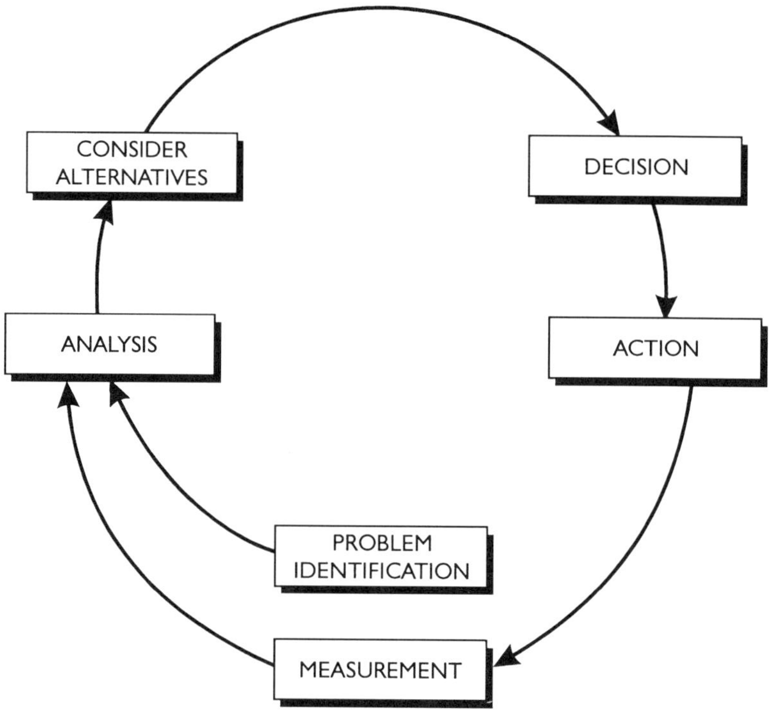

FIGURE 1 *The performance cycle.*

practitioners should consider in resolving safety issues. It depicts a way of thinking about hazards and risks and establishes an effective order of inquiry and action for risk elimination or amelioration. Why propose that safety practitioners adopt a safety decision hierarchy? *The New Rational Manager*, reflecting the real-world observations of its authors, Kepner and Tregoe, makes the case:

> The most effective managers, from the announcement of a problem until its resolution, appeared to follow a clear formula in both the orderly sequence and the quality of their questions and actions (vii).

It makes sense to apply a safety decision hierarchy with an orderly sequence of effective actions to resolve safety issues.

ON PROBLEM IDENTIFICATION AND ANALYSIS

What is the substance of problem identification and analysis for safety practitioners? These excerpts are taken from the author's *On the Practice of Safety, 2nd ed.*:

A. Problem Identification and Analysis

1. Identify and analyze hazards
2. Assess the risks

B. Consider These Actions, in Their Order of Effectiveness

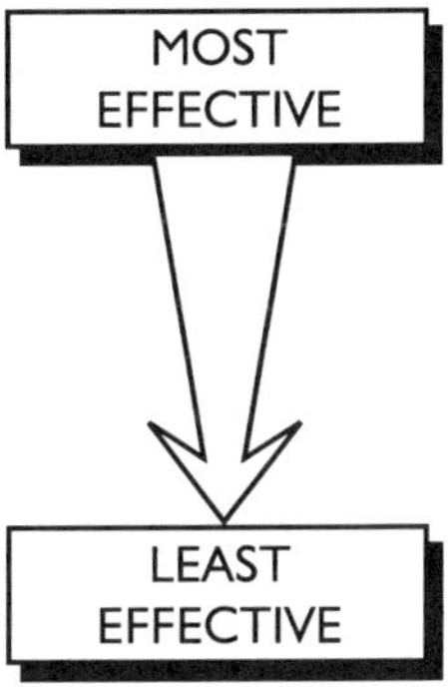

1. Eliminate hazards and risks through system design and redesign
2. Reduce risks by substituting less hazardous methods or materials
3. Incorporate safety devices
4. Provide warning systems
5. Apply administrative controls (work methods, training, etc.)
6. Provide personal protective equipment

C. Decide and Take Action

D. Measure for Effectiveness: Re-Analyze as Needed

FIGURE 2 *The safety decision hierarchy.*

> Hazards are the justification for the existence of safety practitioners. If there are no hazards, if there is no potential for harm, there would be no risk, and safety practitioners would not exist.
>
> The dual nature of hazards must be understood. Hazards include any aspect of technology or activity that produces risk. A hazard is defined as the potential for harm: Hazards include the characteristics of things and the actions or inactions of people.
>
> Risk is defined as a measure of the probability of a hazards-related incident occurring and the severity of harm or damage that could result.
>
> The entirety of purpose of those accountable for safety, whatever their titles, is to manage their endeavors with respect to hazards so that the risks deriving from those hazards are acceptable (15).

In the literature on problem solving, every model has a problem identification and analysis phase. In applying the safety decision hierarchy, the goal in the problem identification and analysis phase would be to identify and analyze the hazards and assess the risks.

Hazard and risk problems cannot be intelligently addressed until the potentials of hazards are established, and assessments are made of the probability of incidents occurring and the possible severity of their consequences.

THE LOGIC OF TAKING ACTION IN DESCENDING ORDER OF EFFECTIVENESS

In the safety decision hierarchy, alternative risk elimination and amelioration actions are listed in a descending order of effectiveness. The rationale for following the order given must be supported.

Amelioration actions described in the first, second, and third action levels are more effective because they are preventive actions that reduce risk by design measures which eliminate or adequately control hazards. Actions described in the fourth, fifth, and sixth levels are contingent actions and rely greatly on the performance of people.

The terms *preventive* and *contingent* come from Kepner and Tregoe (again, *The New Rational Manager*, 147). Their discussion of "identifying likely causes and preventive actions" in the generally applied problem-solving and decision-making process precisely corresponds to the risk elimination and amelioration concepts put forth here:

> Two kinds of actions are available to anyone conducting a Potential Problem Analysis: preventive actions and contingent actions. The effect of preventive actions is to remove, partially or totally, the likely cause of a potential problem. The effectiveness of a contingent action is to reduce the impact of a problem that cannot be prevented. Preventive actions, if they can be taken, are obviously more efficient than contingent actions.

Comments follow on each of the actions listed in the safety decision hierarchy in descending order of effectiveness:

1. If the hazards are eliminated in the design and redesign processes, risks that derive from those hazards are also eliminated. If there are no hazards, there is no potential for harm and, therefore, no risk. Obviously, hazard elimination is the most effective way to eliminate risk.
2. By substituting less hazardous methods or materials (e.g., using automated material-handling equipment rather than manual material handling, or using a less hazardous degreasing material), risks can be substantively reduced in the operations or product design and redesign process, and reliance on the performance of people is minimized.
3. When safety devices are incorporated in the system or product in the form

of engineering controls, marked risk reduction can be achieved and reliance on the worker or product user is further reduced.

4. Warning signs, labels, and instructions have been partially effective. Warning systems are reactionary. They alert people only after the development of a situation for which a hazard's potential is in the process of being realized (e.g., the smoke alarm in a house). Warning system effectiveness relies considerably on training, the quality of maintenance, and the reactions of people.
5. Administrative controls, only some of which are included here, rely entirely on the performance of people, for whom to err is human. Some administrative controls are the development of appropriate work methods and procedures, personnel selection, training, supervision, motivation, work scheduling, job rotation, scheduled rest periods, maintenance, management of change, investigations, inspections, and behavior modification.
6. The proper use of personal protective equipment relies on an extensive series of supervisory and personnel actions, such as the identification of needed equipment, its selection, fitting, training, inspection, maintenance, etcetera.

Particular comment is needed on the use of the term "warning systems" in item 4 above. In one published hierarchy of control, the designation for this same purpose is warning signs; in another, it is warnings. If "warning" is to be included as one of the risk control methods that must be applied, the entire warning system needs consideration. For example, the NFPA Life Safety Code 101 may require, among other things: smoke and products of combustion detectors; automatic and manual audible and visible alarms; lighted exit signs; designated, alternate and properly lit exit paths; adequate spacing for personnel at the end of the exit path; proper hardware for doors; and emergency power systems.

DECIDING AND TAKING ACTION

All components of the safety decision hierarchy are to be considered when deciding on the appropriate actions to be taken. They can be applied to a facility, process, system, piece of equipment, or a tool in its simplest form. The entire safety decision hierarchy is applicable to all three major facets of the practice of safety:

- In the design processes, or preoperational mode, where the opportunities are greatest and the costs lower, for identifying and analyzing hazards and for their avoidance, elimination, or control
- In the operational mode, where, integrated within a continuous improve-

ment process, hazards are identified and evaluated and eliminated or controlled, before their potentials are realized and hazards-related incidents occur
- Post incident, through investigation of hazards-related incidents to determine and eliminate or control their causal factors

Even after applying the safety decision hierarchy, if an operation proceeds, the risk will never be zero. No matter what actions are taken to reduce risk, some residual risk will exist, although perhaps infinitesimal. If the safety decision hierarchy is used effectively, the residual risks will be acceptable.

MEASURING FOR EFFECTIVENESS: REANALYZING AS NECESSARY

Assuring that the actions taken accomplish what was intended is an integral step in any effective problem-solving technique. For safety management purposes, measuring for effectiveness requires ascertaining that the actions taken have, in reality, reduced the risk to the level expected. In a formal way, follow-up activity would determine that:

- The system worked or did not work
- The problem was resolved, only partially resolved, or never went away
- All hazards were or were not addressed
- The actions taken did or did not create new hazards

If the residual risk is not acceptable, the thought process advanced in the safety decision hierarchy would again be applied, commencing with hazard identification and analysis.

For incidents that have a high probability of occurrence, the evidence will be forthcoming soon indicating that the actions taken to reduce risk did or did not do what was expected. If incidents that were to be avoided continue to occur, or if other unexpected types of incidents occur, those events indicate clearly that further analysis and action are needed. For incidents that have a low probability of occurrence, and which too often result in severe injury, an immediate indication will not be forthcoming that the actions taken were successful or not. Incidents that have a low probability of occurrence and severe injury potential deserve a particularly close look in applying the safety decision hierarchy.

SIGNIFICANCE OF THE SAFETY DECISION HIERARCHY

Looking to the future, the application of a safety decision hierarchy will become more prevalent in the practice of safety. The following excerpt is from a 1999 letter of understanding between Daimler Chrysler Corporation and the International Union, UAW, indicating that a hierarchy of controls is to be applied in the selection of risk control methods:

> Risk assessment and hazard control will be conducted for high hazard jobs. This includes identification of tasks, identification of hazards associated with each task, risk assessment and selection of method(s) to control. Selection of controls will be based on the hierarchy, which gives preference to engineering solutions over procedures and personal protective equipment.

Note that hazards are to be identified, risks are to be assessed, and methods of control are to be based on a hierarchy. This is but one example of the importance that the application of a hierarchy of considerations in an order of effectiveness has assumed with respect to hazards and risks.

ORIGINS OF THE SAFETY DECISION HIERARCHY

References to a hierarchy of controls appear in safety literature dating back to at least 1931. Although many safety practitioners have contributed to the emergence of a safety decision hierarchy, all cannot be recognized here. Instead, we will give a brief review of how the safety decision hierarchy evolved, referencing:

- The work of H. W. Heinrich
- The National Safety Council
- The developers of the system safety requirements for providers of materials or services to the federal government
- William Haddon's theory concerning the control of unwanted releases of energy, expanded to include exposures to hazardous environments

H. W. Heinrich's Early Work

Heinrich was a pioneer in the field of accident prevention. In the second edition of *Industrial Accident Prevention: A Scientific Approach*, published in 1941, he introduced a method to deal with occupational diseases. The following excerpt from the 3rd edition of his legendary work, published in 1950, is an extension of his earlier thinking:

> In the case of occupational disease (methods of control) may be expressed as follows:
>
> **Engineering Revision**
>
> 1. Elimination of the injurious substances or sources.
> 2. Reduction of the original amounts or volumes or frequency of use of injurious substances or sources.
> 3. Removal of injurious substances or sources after use.

4. Isolation, guarding, or enclosing of the injurious substances or sources.
5. Provision of personal protective devices.

Personal Approach

6. Control of unsafe personal acts through instruction, persuasion and appeal, personnel adjustment, and discipline (309).

Heinrich uses the terms engineering revision, elimination, reduction, removal, provision of personal protective devices, and instruction: All of these have parallels in the latest versions of the hierarchy of controls. Heinrich's work was truly influential with respect to the control of occupational disease.

The National Safety Council

The 3rd edition of the Council's *Accident Prevention Manual* was published in 1955. Section 4 entitled, "Removing the Hazard From the Job" (4-1), provides guidance similar to that proposed by Heinrich on an order of effectiveness and preference:

> The engineer should include in his planning and follow-through such measures as will attain one of the accident prevention goals listed as follows (in the order of effectiveness and preference):
>
> 1. Elimination of the hazard from the machine, method, material, or plant structure.
> 2. Guarding or otherwise minimizing the hazard at its source if the hazard cannot be eliminated.
> 3. Guarding the person of the operator through the use of personal protective equipment if the hazard cannot be eliminated or guarded at its source.
>
> Company policies should be such that safety can be designed and built into the job, rather than added after the job has been put into operation.

Establishing an order of effectiveness and preference was an important step in the evolution of the safety decision hierarchy. It implies that some steps in the process are preferable since they achieve greater risk reduction than others. Declaring that safety policies should require that safety be designed and built into the job rather than added on later is also a premise that influenced later versions of hierarchies of control.

System Safety Requirements for Federal Government Contractors

In the literature on system safety, some writers trace the principles embodied in Military Standard System Safety Requirements (MIL-STD-882) to the work of aviation personnel in the post-World War II era. This standard was first issued in 1969; the latest revision of it, MIL-STD-882D, was published in 2000. Especially over the last 25 years, the standard has been influential since it expanded on previous writings suggesting that a hierarchy of controls should be applied.

The standard establishes the "design order of precedence for mitigating identified hazards." Keep in mind that it applies to the safety of products, equipment, materials, and services supplied by companies under contract to the federal government. It does not apply to the contractor's operations.

The Design Order of Precedence—System Safety

Eliminate hazards through design selection. If unable to eliminate an identified hazard, reduce the associated mishap risk to an acceptable level through design selection.

Incorporate safety devices. If unable to eliminate the hazard through design selection, reduce the mishap risk to an acceptable level through using protective safety features or devices.

Provide warning devices. If safety devices do not adequately lower the mishap risk of the hazard, include a detection and warning system to alert personnel to the particular hazard.

Develop procedures and training. Where it is impractical to eliminate hazards through design selection or to reduce the associated risk to an acceptable level with safety and warning devices, incorporate special procedures and training. Procedures may include the use of personal protective equipment. For hazards assigned Catastrophic or Critical mishap severity categories, avoid using warning, caution or other written advisory as the only risk reduction method.

Precedence means priority in order, rank, or importance. For many risk situations, a combination of the risk management methods shown in the order of precedence will apply. But, the expectation is that consideration will be given to each of the steps in the order of precedence in descending order, and that reasonable attempts will be made to eliminate hazards or reduce their associated risks through steps higher in the hierarchy before lower steps are considered. A lower step in the order of precedence is not to be chosen until practical applications of the preceding higher level or levels are exhausted.

Haddon's Unwanted Energy Release Concept

William Haddon, the originator of the energy release theory, was the first director of the National Highway Safety Bureau. His work is considered seminal because it influenced how many safety practitioners and authors think about risk elimination or amelioration. Haddon's concept is that unwanted transfers of energy can be harmful (and wasteful) and that a systematic approach to limiting their possibility should be adopted.

Although Haddon stated in "On the Escape of Tigers: An Ecologic Note" that "the concern here is the reduction of damage produced by energy transfer," he also said that "the type of categorization here is similar to those useful for dealing systematically with other environmental problems and their ecology." The following is an excerpt are from Haddon's break-through paper:

> A major class of ecologic phenomena involves the transfer of energy in such ways and amounts, and at such rapid rates, that inanimate or animate structures are damaged.
>
> Several strategies, in one mix or another, are available for reducing the human and economic losses that make this class of phenomena of social concern. In their logical sequence, they are as follows:
>
> - prevent the marshalling of the form of energy;
> - reduce the amount of energy marshaled;
> - prevent the release of the energy;
> - modify the rate or spatial distribution of release of the energy from its source;
> - separate, in space or time, the energy being released from that which is susceptible to harm or damage;
> - separate, by interposing a material barrier (the energy released from that which is susceptible to harm or damage);
> - modify appropriately the contact surface, subsurface, or basic structure, as in eliminating, rounding, and softening corners, edges, and points with which people can, and therefore sooner or later do, come in contact;
> - strengthen the structure, living or non-living, that might otherwise be damaged by the energy transfer;
> - move rapidly in detection and evaluation of damage that has occurred or is occurring, and counter its continuation or extension; and
> - after the emergency period following the damaging energy exchange, stabilize the process.

Unfortunately, all hazards are not addressed by the unwanted energy release concept. Such examples are the potential for asphyxiation from entering a confined space filled with inert gas, or inhalation of asbestos fibers. But, all hazards are encompassed within this single goal: avoiding both unwanted energy releases and exposures to hazardous environments.

A GUIDELINE

To provide guidance to safety practitioners who are contemplating the application of the safety decision hierarchy, we will reproduce a guideline entitled, "General Design Requirements: A Thought Process for Hazard Avoidance, Elimination, or Control," which originally appeared in the author's *On the Practice of Safety* (162). This guideline is an extension of the incident prevention aspects of Haddon's work, but it also reflects the references previously made to Heinrich, the National Safety Council, and system safety program requirements.

The guideline addresses nine major subjects. Haddon himself listed ten strategies (one of which is divided into two parts here). And, his last two subjects pertain to recovery actions after an incident occurs, which are not addressed here. In no way is it suggested that the author's guideline considers all hazard and risk elimination or amelioration possibilities. It can be helpful as a reference.

General Design Requirements: A Thought Process for Hazard Avoidance, Elimination, or Control

1. Avoid introduction of the hazard: prevent buildup of the form of energy or hazardous materials
   - avoid producing or manufacturing the energy or the hazardous material
   - use material handling equipment rather than manual means
   - don't elevate persons or objects
2. Limit the amount of energy or hazardous material
   - seek ways to reduce actual or potential energy input
   - use the minimum energy or material for the task (voltage, pressure, chemicals, fuel storage, heights)
   - consider smaller weights in material handling
   - store hazardous materials in smaller containers
   - remove unneeded objects from overhead surfaces
3. Substitute, using the less hazardous:
   - substitute a safer substance for a more hazardous one: when hazardous materials must be used, select those with the least risk throughout the life cycle of the system
   - replace hazardous operations with less hazardous operations
   - use designs needing less maintenance
   - use designs that are easier to maintain, considering human factors
4. Prevent unwanted energy or hazardous material buildup
   - provide appropriate signals and controls
   - use regulators, governors, and limit controls
   - provide the required redundancy
   - control accumulation of dusts, vapors, mists, etc.
   - minimize storage to prevent excessive energy or hazardous material buildup
   - reduce operating speed (processes, equipment, vehicles)
5. Prevent unwanted energy or hazardous material release
   - design containment vessels, structures, elevators, material handling equipment to appropriate safety factors
   - consider the unexpected in the design process, to include avoiding the wrong input
   - protect stored energy and hazardous material from possible shock
   - provide fail-safe interlocks on equipment, doors, valves
   - install railings on elevations
   - provide non-slip working surfaces
   - control traffic to avoid collisions

6. Slow down the release of energy or hazardous material
   - provide safety and bleed off valves
   - reduce the burning rate (using an inhibitor)
   - reduce road grade
   - provide error-forgiving road margins
7. Separate in space or time, or both, the release of energy or hazardous materials from that which is exposed to harm
   - isolate hazardous substances, components, and operations from other activities, areas, and incompatible materials, and from personnel
   - locate equipment so that access during operations, maintenance, repair, or adjustment minimizes personnel exposure (e.g., hazardous chemicals, high voltage, electromagnetic radiation, cutting edges)
   - arrange remote controls for hazardous operations
   - eliminate two-way traffic
   - separate vehicle from pedestrian traffic
   - provide warning systems and time delays
8. Interpose barriers to protect the people, property, or the environment exposed to an unwanted energy or hazardous material release
   - insulation on electrical wiring
   - guards on machines, enclosures, fences
   - shock absorbers
   - personal protective equipment
   - directed venting
   - walls and shields
   - noise controls
   - safety nets
9. Modify the shock concentrating surfaces
   - padding on low overheads
   - rounded corners
   - ergonomically designed tools
   - "soft" areas under playground equipment

If a hazard cannot be eliminated, the decision to ameliorate the risks will more than likely involve a combination of the actions described in the safety decision hierarchy. Decisions should be documented, and the prescribed actions then taken. Also, it should be determined that the prescribed remediation actually occurs.

CONCLUSION

Adopting well-established problem-solving techniques to address hazard and risk problems is a fundamentally sound approach. This chapter offered an effective and practical safety decision hierarchy for risk elimination and amelioration, one that can incorporate good problem-solving and decision-making techniques.

Looking to the future, the author believes that knowledge of the safety decision hierarchy outlined here will be sought by safety practitioners in search of solutions to common concerns.

REFERENCES

Accident Prevention Manual, 3rd ed. Itasca, IL: National Safety Council, 1955.

Bransford, John D. and Barry S. Stein. *The Ideal Problem Solver.* New York: W. H. Freeman and Company, 1984.

Excerpts from the 1999 Letter of Understanding, Daimler Chrysler Corporation and the International Union, UAW. (www.uaw.org/contracts)

Haddon, William J., Jr. "On the Escape of Tigers: An Ecological Note." *Technology Review*, May 1970.

Heinrich, H. W. *Industrial Accident Prevention: A Scientific Approach*, 3rd ed., New York: McGraw-Hill, 1950.

Johnson, William G. *MORT Safety Assurance Systems.* New York: Marcel Dekker, 1980.

Kepner, Charles H. and Benjamin B. Tregoe. *The Rational Manager.* New York: McGraw-Hill, 1965.

Kepner, Charles H. and Benjamin B. Tregoe. *The New Rational Manager.* Princeton, NJ: Princeton Research Press, 1981.

Manuele, Fred A. *On the Practice of Safety*, 2nd ed. New York: John Wiley & Sons, 1997.

"Military Standard System Safety Program Requirements (MIL-STD-882D)." Washington, DC: Dept. of Defense, 2000.

5

Why You Need to Know About Hazard Analysis and Risk Assessment

INTRODUCTION

Safety standards and guidelines issued in recent years or that are in the process of development require or propose *hazard analyses* and *risk assessments* for workplace equipment, machinery, and processes. This is a significant and evolving progression that will have an important bearing on the knowledge needs of safety practitioners and the substance of their practice.

In this chapter, some of the relevant standards and guidelines requiring or proposing hazard analyses and risk assessments will be reviewed. There are similarities and differences in the approaches taken in the existing standards and guidelines, and in the relative work in progress. In their entirety, the developments of recent years make it important for safety practitioners to have knowledge of hazard analysis and risk assessment basics.

A SIGNIFICANT EVENT

A noteworthy first was achieved in the United States in June 1999 when approval was given for ANSI/RIA R15.06-1999, which is the American National Standard for Industrial Robots and Robot Systems—Safety Requirements. The Robotic Industries Association is the secretariat for the standard.

Why should generalist safety practitioners pay attention to this standard? This is the first occupational safety standard issued by the American National Standards Institute (ANSI) in which a comprehensive risk assessment is presented as a means to determine the design requirements, and to apply the safeguarding and subsequent administrative controls that may be needed. This standard affects both the makers and users of robots and robot systems.

The robotic standard requires that a strategy be developed for identifying and controlling hazards and safeguarding personnel by either applying the specifically prescribed guarding methods set forth in the standard or by conducting a comprehensive risk assessment to determine the amount or kind of safeguarding necessary.

The standard is thorough in terms of task and hazard identification, risk estimation, determination of risk-reduction categories in relation to the estimated risk, and the minimum required safeguard performance for each risk-reduction category. For example, if it is determined through a risk assessment that for the risk-reduction category defined, serious injury could occur, and the exposure is frequent, and worker avoidance of the injury is unlikely, the safeguard performance to be considered first is hazard elimination or substitution.

Annex C, which is informative but not part of the actual standard, includes "an example of a risk assessment method which has been demonstrated to achieve the desired results." However, any risk-assessment method is acceptable that prescribes safeguarding equivalent to the requirements of the standard.

To assist users of the standard in making risk assessments, a Robot Risk Assessment CD is available from the Robotics Industries Association. The promotional literature states the following:

> The Robot Risk Assessment CD walks you through the stages of a risk assessment following the guidelines set in the new safety standard. This easy to navigate program helps you conduct a hazard analysis based on the application of the associated tasks the robot will be completing. It then leads you through the risk assessment to determine the severity of potential injuries and the frequency of exposure and likelihood of avoidance. Once the risk assessment is complete, the program provides for easy documentation.

Including risk assessment provisions in this ANSI standard could very well be an indicator that similar provisions will be integrated into other standards. This further suggests that safety practitioners whose work is influenced by ANSI standards should become proficient in hazard analysis and risk assessment.

A MAJOR EVENT IN HAZARDS ANALYSIS AND RISK ASSESSMENT: B11.TR3

TR3 is the acronym used by the B11.TR3 Subcommittee formed by ANSI's Machine Tool Safety Standards Committee (B11). TR stands for technical report. The committee's work is entitled, "Risk assessment and reduction—A guideline to estimate, evaluate and reduce risks associated with machine tools." The secretariat for this work is the Association for Manufacturing Technology. TR3 became a registered document at ANSI in November 2000 and is available through the secretariat.

The following are adaptations of the principles guiding the work of the TR3 subcommittee:

- A simple, practical, and generic hazard analysis and risk assessment process is to be developed that has the potential to be incorporated in all B11 standards.
- The process must apply to both suppliers and users.
- To the extent possible, the technical report is to be harmonized with European Standards.

This technical report is one of the ANSI B11 series of reports and standards pertaining to the design, construction, care and use of machine tools. It is a guideline, not a standard. However, if the content of the guideline is incorporated into all B11 standards, its impact on the design, manufacture, and use of machinery will be immense. Discussions by the committees for individual machine tools on including the principles of TR3 in their standards have commenced.

More important, over 90% of the guideline is generic. Thus, it is a basic document on hazard analysis and risk assessment providing guidance on reducing risks according to a prioritized procedure and on the selection of appropriate design and protective measures. When the process is complete, a tolerable risk level will be achieved.

It is significant that the TR3 document addresses the safety responsibilities of both designers and manufacturers, and of purchasers and users. Now that the TR3 report has been adopted by ANSI, it is logical to assume that other business and industry segments will be influenced by the work.

WORK IN PROGRESS WITH THE POTENTIAL FOR BROAD IMPACT

An accredited standards committee is working on a revision of ANSI Z244 entitled, "Control of Hazardous Energy—Lockout/Tagout and Alternative Methods." Its October 2000 draft requires that a risk assessment be performed to determine the need for and design sufficiency of appropriate energy isolating devices and systems.

A section on risk assessment outlines in some detail the procedures to be followed in making a qualitative appraisal of the risks (see Chapter 6). Also, a hierarchy of controls is presented that is comparable to Table 1 in Chapter 4.

With respect to the control of hazardous energy and lockout/tagout systems and procedures, responsibilities are outlined for manufacturers, integrators, modifiers, remanufacturers, and users. Revisions in this standard will impact on a very large number of American workplaces.

This revision of an important ANSI standard is another indication of a trend that emphasizes the need for safety practitioners to acquire a knowledge of hazard analysis and risk assessment techniques.

EUROPEAN INFLUENCES

Standards adopted by the European Community for Standardization governing the safety of machinery that goes into the workplace have had considerable impact on American manufacturers who export machinery to Europe. These standards require that hazards analyses and risk assessments be made in the design and manufacturing processes.

How broad is such an impact? At a presentation during the 1999 National Safety Congress, one speaker commented that 100% of the equipment made by her company met the European Community standards. It has also been said that one of the world's largest companies has adopted a policy whereby only by exception would equipment for any use be purchased that did not meet those requirements. Trade associations effected by the European standards have written applicable guidelines for their industries and are conducting training programs.

Many companies now have designated personnel within their staff who establish procedures for meeting the safety requirements in design and manufacturing processes as outlined in European standards. Also, there are a growing number of consultants offering advice to companies on meeting these standards.

There are several European Committee standards governing the safety of equipment that will exist in the workplace. Two are particularly pertinent: EN 292 and EN 1050.

Standard EN 292 has two parts. Part 1 is entitled, "Safety of machinery—Basic concepts, general principles for design." It covers basic terminology and methodology. Part 2 is entitled, "Technical principles and specifications." This is the scope presented in Part 1:

> This European standard defines basic terminology and specifies general design methods, to help designers and manufacturers in achieving safety in the design of machinery for professional and non-professional purposes. It may also be used for other technical products having similar hazards.

It is specified in Part 1 that the designer must complete, in all circumstances, the following activities in the order given:

- Specify the limits of the machine
- Identify the hazards and assess the risks
- Remove the hazards or limit the risks as much as possible
- Design guards and/or safety devices (safeguards) against any remaining risks
- Inform and warn the user about any residual risks.
- Consider any necessary additional precautions.

Since EN 292 required that hazards be identified and risks assessed, standard EN 1050 entitled, "Safety of machinery—Principles for risk assessment" was developed. From its Scope section:

> This standard gives guidance on the information required to allow risk assessments to be carried out. Procedures are described for identifying hazards and estimating and evaluating risk. The purpose of the standard is to provide advice for decisions to be made on the safety of machinery and the type of documentation required to verify the risk assessment carried out.

EN 1050 does what it purports to do: It enumerates a hazard identification and risk assessment methodology.

As of 1997, most products sold in the European economic community are required to bear a "CE" mark. To obtain the classification, a "responsible person" (most often the manufacturer) must ensure that certain documentation concerning the safety of a product is available for inspection by enforcement authorities. By affixing the CE mark, a manufacturer indicates that the product complies with all applicable directives. The CE mark is the passport for goods entering European Union countries.

It is important to understand that the EN 292 and EN 1050 standards exist to help designers and manufacturers in achieving safety in the design of machinery. These standards apply to designers and manufacturers of machinery only. They do not apply to purchasers and users of equipment.

There are no standards in the United States having the force of law that require designers or manufacturers of machinery and equipment to document that hazards analyses and risk assessments have been made, and that appropriate prevention and control methods have been undertaken.

DEVELOPMENTS IN AVIATION GROUND SAFETY

One of the most interesting innovations regarding hazard analysis and risk assessment was achieved by the International Air Transport Section of the National Safety Council. The section is truly international, including representatives from all the continents. Its *Aviation Ground Operation Safety Handbook* is used throughout the world (a 5th edition was published in July 2000).

New material entitled, "Risk Management Guide for the Aviation Industry" was added to the *Handbook*; which, to emphasize, pertains to ground operations. The following is an excerpt from the introduction to that material:

> Risk management takes aviation safety to the next level. It is a six-step, logic-based, common-sense approach to making calculated decisions on human, material, and environmental factors before, during and after operations. Risk management enables senior leaders, functional managers, supervisors and individuals to maximize opportunities for success while minimizing risks.

TABLE 1 Risk Management Guide for the Aviation Industry. Aviation Ground Operation Safety Handbook

1. Identify the hazard
2. Assess the risk
3. Analyze risk control measures
4. Make control decisions
5. Implement risk controls
6. Supervise and review

The air transport group has outlined a way of thinking about and dealing with hazards and risks, applying a logical and sequential methodology. They have developed a "process to detect, assess, and control risk." The highlights of their "six-step, logic-based, common-sense approach" are given in Table 1. The discussion of each step in the Handbook is extensive.

We will address the first two of these six steps here. The tools listed as aids to be used in identifying the hazards, as in step 1, follow in Table 2. For step 2 in Table 1, assess the risks, a risk-assessment matrix is available in the Handbook from which the user can develop risk levels based on the probability of a hazardous event occurring and the severity of its consequences. (More about risk assessment matrices in the next chapter.)

TABLE 2 Tools to Aid in Hazard Identification. Aviation Ground Operation Safety Handbook

1. *Operations Analysis*: To understand the flow of events
2. *Preliminary Hazard Analysis (PHA)*: To conduct a quick survey of all phases of an operation
3. *"What-If" Tool*: To capture the input of personnel in a brainstorming environment
4. *Scenario Process Tool*: To use one's imagination and visualizations to capture hazards
5. *Change Analysis*: To detect the hazard implications of both planned and unplanned change

Knowing how to use the listed hazard identification and analysis methods will be necessary for those who wish to apply the concepts outlined in the "Risk Management Guide." This work by the air transport safety professionals is innovative and shows leadership capability. They saw a need and fulfilled it. Others should be attentive to its content.

THE SEMICONDUCTOR INDUSTRY

Guidelines issued by the semiconductor industry provide another indication of a trade association recognizing the value of hazards analyses and risk assessments to eliminate or control hazards and to attain acceptable risk levels. To do what issued guidelines propose requires the involvement of personnel at several levels who are skilled in hazard analysis and risk assessment.

The manufacturers of machinery and equipment used in making semiconductors and their customers (e.g., Intel and IBM) recognized that their mutual interests would be well served if that equipment met agreed upon safety guidelines. Their trade association, The Semiconductor Equipment and Materials International, has issued several safety-related guidelines, two of which are of particular interest: SEMI S2 and SEMI S10.

SEMI S2-0200 is entitled, "Environmental, Health, and Safety Guideline For Semiconductor Manufacturing Equipment." This guideline and the European Community standards discussed previously have a similarity of purpose and limitation. They pertain to manufacturers of machinery and equipment that will go into a workplace; they do not pertain to the users of equipment. The following is the stated purpose of S2:

> This safety guideline is intended as a set of performance-based environmental, health, and safety (EHS) considerations for semiconductor and flat panel display manufacturing equipment.

It applies principally to suppliers and what they are to do with respect to hazards and risks in the design and manufacturing process. As the guideline clearly states,

> the assumption is made that operators, maintenance personnel, and service personnel are trained in the tasks that they are intended to perform.

Suppliers are advised that a hazard analysis should be performed early in the design phase to identify and evaluate hazards, and that the analysis should be updated as the design matures. A supplementary and advisory document to S2 is entitled, "Related Information 1—Equipment/Product Safety Program." It makes an interesting statement about the need, sometimes, to go beyond issued safety standards in the design process:

> Compliance with design-based safety standards does not necessarily ensure adequate safety in complex or state-of-the-art systems. It often is necessary to perform hazard analyses to identify hazards that are specific with the system, and develop hazard control measures that adequately control the associated risk beyond those that are covered in existing design-based standards.

Risks deriving from the hazards identified and evaluated through the process outlined in S2 are to be characterized using SEMI S10, "Safety Guideline for Risk Assessment." The following is the stated purpose of S10:

> This document provides a method for assessing the risk associated with any hazard and for ranking different hazards according to the risks they present. The method can be used to differentiate between those hazards which are acceptable and others which require further mitigation to reduce their risk to acceptable levels.

The suggested procedures require that, in the risk assessment process, both the expected severity of consequences of an incident resulting from a hazard's potential being realized and the probability of its occurrence be evaluated. Examples of mishap severity and likelihood groups are given.

The guideline then indicates in a logical sequence that after the hazard analysis and risk assessment are concluded, decisions are to be made about whether the risk is acceptable or further mitigation is necessary. As the guideline clearly states,

> the Risk Assessment Matrix (such as the example shown) is a tool for making this judgment in a consistent, repeatable way.

Guidelines issued by Semiconductor Equipment and Materials International are truly international. Many countries were represented on its guideline committees. This organization's work has had considerable influence on safety in the semiconductor industry.

THE CHEMICAL INDUSTRY: OSHA REQUIREMENTS

"OSHA's Rule For Process Safety Management Of Highly Hazardous Chemicals" applies to employers at about 50,000 locations, many of which are not chemical companies. As an indicator of the trend toward including requirements for hazards analyses in guidelines and standards, this Occupational Safety and Health Administration (OSHA) standard merits a review by safety practitioners. It requires the following:

> The employer shall perform an initial hazard analysis (hazard evaluation) on processes covered by this standard. The process hazard analysis shall be appropriate to the complexity of the process and shall identify, evaluate, and control the hazards involved in the process.

The employer shall use one or more of the following methodologies that are appropriate to determine and evaluate the hazards of the process being analyzed:

- What-If;
- Checklist;
- What-If/Checklist;
- Hazard and Operability Study (HAZOP);
- Failure Modes and Effect Analysis (FMEA);
- Fault Tree Analysis; or
- An appropriate equivalent methodology.

Also, the hazard analysis shall address:

- The hazards of the process;
- The identification of any previous incident which had a likely potential for catastrophic consequences in the workplace;

- Engineering and administrative controls applicable to the hazards and their interrelationships;
- Consequences of failure of engineering and administrative controls;
- Facility citing;
- Human factors; and
- A qualitative evaluation of a range of the possible safety and health effects of failure of controls on employees in the workplace.

Under the requirements for *pre-startup safety review*, the employer is required to provide, for new facilities and for significant modifications, a process hazard analysis—among other considerations.

In no place in the standard is there mention of occurrence probability. This appears in the preamble to the standard:

> OSHA has modified the paragraph (editorial note—paragraph on consequence analysis) to indicate that it did not intend employers to conduct probabilistic risk assessments to satisfy the requirement to perform a consequence analysis.

However, all risks are not equal. And, managements do consider incident probability in their decision making, recorded or not recorded, when determining the order of resource allocation to individual projects.

This OSHA standard may be an important indicator of things to come regarding requirements for hazards analyses and risk assessments.

THE CHEMICAL INDUSTRY: EPA REQUIREMENTS

The Environmental Protection Agency (EPA) and OSHA have different legal authority with respect to the accidental release of harmful substances. The concerns at EPA center on offsite consequences: that is, harm to the public and the environment. At OSHA, the legal authority pertains to on-site consequences.

On August 19, 1996, EPA issued Rule 40 CFR Part 68, "Risk Management Programs for Chemical Accidental Release Prevention." Risk Management Plans required of location managements by the rule were due by June 21, 1999. Although the provisions of the rule are extensive, only the specifications for hazards analyses will be addressed here.

Processes subject to this rule are divided into three groups: labeled by EPA as Programs 1, 2, and 3. Program levels relate to the quantities and extent of exposure to toxic and flammable chemicals.

For locations qualifying for Program levels 1 and 2, those with less exposure, EPA will accept hazard reviews done by qualified personnel using suitable checklists. Hazard reviews must be documented and show that problems have been addressed. EPA makes reference in its literature to the desirability of using the "What If" hazard identification and analysis process, bringing personnel into discussions of what could occur, to extend the value of checklists. EPA also proposes the use of more involved analytical techniques if findings suggest that to be desirable.

The hazard review requirement for Program level 3 locations are more specific and extensive. But, those locations that are compliant with the OSHA rule for the process safety management of highly hazardous chemicals will need to do little new, although they will need to extend their hazards analyses to consider the probability of harm to the public or the environment. As with OSHA, a team must complete the process hazards analyses required by EPA, one member of which has had prior experience with the process.

As would be expected at locations with more significant exposures, the process hazard analysis requirements are more extensive. They must be documented and include:

- Hazards of the process
- Identification of previous, potentially catastrophic incidents
- Engineering and administrative controls applicable to the hazards
- Siting
- Human factors
- Qualitative evaluation of health and safety impacts of control failure

One or more analytical techniques must be used in making process hazards analyses. The list of such techniques in the EPA literature is identical to that in OSHA's "Process Safety Management" standard. Those techniques are included in the comments previously made here on the OSHA standard. As is the case with OSHA, probability of occurrence estimates need not be part of hazards analyses.

For American industry, EPA has obviously extended the knowledge and skill requirements regarding hazard analysis techniques.

THE CHEMICAL INDUSTRY: THE EXTENSIVE BODY OF INFORMATION

Completing hazards analyses was a common practice in the chemical industry many years before requirements for their use were established by OSHA and EPA. Although that practice is not of recent origin, it is mentioned here because of its extensive knowledge requirements.

The body of information available in the chemical industry on hazard analysis is extensive. But, reference will be made here to only one publication because of its particular significant. The Center For Chemical Process Safety, a part of the American Institute of Chemical Engineers, published *Guidelines For Hazard Evaluation Procedures, Second Edition with Worked Examples* in 1992. The fact that the text was written by a chemically oriented organization should not dissuade those who seek an education in the following evaluation techniques:

- Safety Review
- Checklist Analysis
- Relative Ranking
- Preliminary Hazard Analysis
- What-If analysis
- What-If Checklist Analysis
- Hazard and Operability Analysis
- Fault Tree Analysis
- Event Tree Analysis
- Cause-Consequence Analysis
- Human Reliability Analysis

These techniques are addressed broadly in the *Guidelines*, within chapters entitled, "Overview of Hazard Evaluation Techniques" and "Using Hazard Evaluation Techniques." Brief descriptions of some of these techniques will be given in the following chapter.

CONCLUSION

Other than for a few industries, incorporating hazard analysis and risk assessment methods in safety standards and guidelines is a recent development. It is logical to assume that this trend will continue and that more safety practitioners will be expected to acquire the necessary related skills. Those who are astute will do so in anticipation of the need.

Further, it should be a goal of safety professionals to have hazards analyses and risk assessments made for all equipment that is to be brought into a workplace. Looking ahead several years, this may regularly occur as the recognition grows that it is good business and good safety management to avoid bringing hazards into the work environment. And, safety professionals should be leading participants in such developments.

REFERENCES

"American National Standard for Industrial Robots and Robot Systems—Safety Requirements." ANSI/RIA R15.06—1999, Ann Arbor, MI: Robotics Industries Association, 1999.

Aviation Ground Operation Safety Handbook, 5th ed. Itasca, IL: National Safety Council, 2000.

"Environmental, Health, and Safety Guidelines for Semiconductor Manufacturing Equipment." SEMI S2-00, Mountain View, CA: Semiconductor Equipment and Materials International, 2001.

Guidelines for Hazard Evaluation Procedures, Second Edition with Worked Examples. New York: Center for Chemical Process Safety of the American Institute of Chemical Engineers, 1992.

"OSHA's Rule for Process Safety Management of Highly Hazardous Chemicals." 1910.119 of 29 CFR 1910, Washington, DC: Dept. of Labor, OSHA, 1992.

Risk Assessment and Reduction—A Guideline to Estimate, Evaluate and Reduce Risks Associated with Machine Tools. McLean, VA: The Association for Manufacturing Technology, 2000.

"Risk Management Programs for Chemical Accidental Release Prevention." 40 CFR Part 68. Washington, DC: EPA, 1996.

"Safety Guidelines for Risk Assessment." SEMI S10-1296, Mountain View, CA: Semiconductor Equipment and Materials International, 1996.

"Safety of Machinery—Basic Concepts, General Principles for Design; Part 1. Basic Terminology, Methodology." European Standard EN 292-1, Brussels: Central Secretariat, European Committee for Standardization, 1991.

"Safety of Machinery—Basic Concepts, General Principles for Design; Part 2. Technical Principles and Specifications." European Standard EN 292-2, Brussels: Central Secretariat, European Committee for Standardization, 1991 (amended 1997).

"Safety of Machinery—Principles for Risk Assessment." European Standard EN 1050, Brussels: Central Secretariat, European Committee for Standardization, 1997.

6

A Primer on Hazard Analysis and Risk Assessment

INTRODUCTION

Since the previous chapter established that safety practitioners need to know about hazard analysis and risk assessment, the next question logically follows: What do they need to know? Literature on the subject is extensive, and a comprehensive reading list appears at the end of this chapter.

The intent here is to provide a primer on hazard analysis and risk assessment that will serve many of the needs of safety practitioners, but not necessarily all. This chapter will:

- Establish what a hazard analysis is
- Discuss how a hazard analysis is extended into a risk assessment
- Outline the steps to be followed in conducting a hazard analysis and risk assessment
- Include descriptions of the most commonly used techniques
- Present examples of basic, two-dimensional risk decision matrices

DEFINING HAZARD, HAZARD ANALYSIS, RISK, AND RISK ASSESSMENT

When a safety practitioner identifies a hazard and its potential for harm or damage and assumes that there is a probability that an injurious or damaging incident can occur, that person has subjectively made a risk assessment. In doing so, for the simpler and less complex hazards and risks, that assessment may be based entirely on *a priori* knowledge and experience, without extensive documentation. Making informal risk assessments has been an integral part of the practice of safety from time immemorial.

Recent developments take safety to a higher level by formalizing the hazard analysis and risk assessment processes. This results in a better appreciation of the significance of individual risks and allows prioritizing them so that more intelligent management decisions can be made with respect to their resolution. As this evolution proceeds, it would be well to have an understanding of the nature of hazards, risks, hazards analyses, and risk assessments. A rather simple but entirely adequate definition of hazards has evolved:

> A hazard is the potential for harm. If there is no potential for harm, injury or damage cannot occur.
>
> Hazards include the characteristics of things and the actions or inactions of people. Hazards encompass all aspects of technology or activity that produce risk.

A hazard analysis establishes the potential for and the probable extent of harm or damage, but need not include an estimate of event probability. Examples of hazards analyses that do not incorporate probability indicators are the estimates of *maximum foreseeable loss* or *maximum probable loss* made by fire protection engineers for insurance purposes, and the hazard analysis requirements of "OSHA's Rule For Process Safety Management Of Highly Hazardous Chemicals, 1910.119."

Whether hazardous situations are simple or complex, completing a hazard analysis is a process that addresses the following questions:

- Is there potential for harm, deriving from the characteristics of things or the actions or inactions of people?
- Can the potential be realized?
- Who and what are exposed to harm or damage?
- What is the frequency of endangerment?
- What will the consequences be, i.e., the severity of harm or damage if the potential is realized?

Conducting a hazard analysis is necessary to and precedes making a risk assessment:

> Risk is defined as a measure of the probability of a hazards-related incident occurring and the severity of harm or damage that could result.

In an appropriate statement indicating the level of risk, both the probability of occurrence and severity of outcome must be included. After determining the severity of expected damage or harm through a hazard analysis, estimating the probability of an incident occurring is the additional and necessary step in concluding a risk assessment.

This excerpt from the *Framework for Environmental Health Risk Management* issued by the Presidential/Congressional Commission on Risk Assessment and Risk Management is an indication of the widespread adoption of these definitions:

> **What Is "Risk"**
>
> Risk is defined as the probability that a substance or situation will produce harm under specified conditions. Risk is a combination of two factors:
>
> - The **probability** that an adverse event will occur;
> - The **consequences** of the adverse event.
>
> Risk encompasses impacts on public health and on the environment, and arises from **exposure** and **hazard**. Risk does not exist if exposure to a harmful substance or situation does not or will not occur. Hazard is determined by whether a particular substance or situation has the potential to cause harmful effects.

Johnson said this about hazard analysis in *MORT Safety Assurance Systems:*

> Hazard analysis is the most important safety process in that, if that fails, all other processes are likely to be ineffective (245).

Johnson's premise is sound: It is one of the most significant fundamentals in the practice of safety and will be elaborated on here.

CONDUCTING A HAZARD ANALYSIS/RISK ASSESSMENT

For many hazards and the risks that derive from them, knowledge gained by safety practitioners through education and experience will lead to proper conclusions on how to attain an acceptable risk level, without bringing together teams of people for discussion. For the more complex situations, it is vital to seek the counsel of experienced personnel who are close to the work or process.

Reaching group consensus on the judgments made would be the goal. Sometimes, for what the safety practitioner considers obvious, having a discussion and achieving consensus are still desirable so that agreement on the actions to be taken and the expenditures to be made is obtained. A general guide follows on how to make a hazard analysis and conclude it with a risk assessment.

Whatever the simplicity or complexity of the hazard/risk situation, and whatever the analysis method used, the following thought and action process is applicable.

1. Establish the Analysis Parameters

Select a manageable task, system, process, or product to be analyzed, establish its boundaries and operating phase (standard operation, maintenance, startup), and define its interface with other tasks or systems, if appropriate. Determine the scope of the analysis in terms of what can be harmed or damaged: people (the public, employees); property; equipment; productivity; the environment.

2. Identify the Hazards

A mindset should be adopted that gets to the bases of causal factors, which are hazards. The following question should be asked: What are the characteristics of things or the actions or inactions of people that present a potential for harm?

Depending on the complexity of the hazardous situation, some or all of the following may apply:

- Use intuitive engineering and operational sense: This is paramount throughout.
- Examine system specifications and expectations.
- Review codes, regulations, and consensus standards.
- Interview current or intended system users or operators.
- Consult checklists.
- Review studies from other similar systems.
- Consider the potential for unwanted energy releases and exposures to hazardous environments.
- Review the historical data: industry experience, incident investigation reports, OSHA and National Safety Council data, manufacturer's literature.
- Brainstorm.

3. Consider the Failure Modes

Define the possible failure modes that would result in a realization of the potentials of hazards. Ask—how could the undesirable event happen.

4. Determine the Frequency and Duration of Exposure

For each harm or damage category selected in step 1 for the scope of the analysis (people, property, etc.), estimate the frequency and duration of exposure to the hazard: i.e., the frequency and duration of endangerment. This is a very important part of this exercise. More judgments will be made during this process than one might realize.

5. Assess the Severity of Consequences

Learned speculations are to be made on the consequences of an occurrence: the number of fatalities, injuries, or illnesses; the value of property or equipment damaged; the time for which productivity will be lost; and the extent of environmental damage. Historical data can be of great value as a baseline. On a subjective basis, the goal would be to decide on the worst credible consequences should an incident occur, not the worst conceivable consequence. When the severity of consequences is determined, a hazard analysis will have been completed.

6. Determine Occurrence Probability

Unless empirical data are available, and that would be a rarity, the process of deciding the probability of an incident occurring will again be subjective. For the more complex hazardous scenario, brainstorming with knowledgeable people is advantageous. Probability has to be related to an interval base of some sort, such as a unit of time or activity, events, units produced, or the life cycle of a facility, equipment, process, or product.

7. Define the Risk

Conclude with a statement that addresses both the probability of an incident occurring and the expected severity of adverse results. Categorize each risk as acceptable or unacceptable.

8. Rank Risks in Order of Priority

To properly communicate with decision makers, a risk ranking system should be adopted so that priorities can be established. Since the hazard analysis and risk assessment exercise is subjective, the risk ranking system would also be subjective.

9. Develop Remediation Proposals

When required by the results of the risk assessment, alternate proposals for the design and operational changes necessary to achieve an acceptable risk level would be recommended. In their order of effectiveness, the list of actions given in Chapter 4, "The Safety Decision Hierarchy," would be the basis upon which remedial proposals are formulated. For each proposal, remediation cost would be determined and an estimate given of its effectiveness in achieving risk reduction.

Care should be taken that remedial actions do not introduce new hazards. If that occurs, the risk must be reevaluated and other countermeasures proposed.

Having an accepted and well-understood risk assessment matrix makes reaching a consensus easier to assess the severity of consequences, determine event probability, define the risk, and rank risks in order of priority.

> Although a hazard analysis and a risk assessment would result from applying the preceding outline, good management requires that the remaining steps in the Safety Decision Hierarchy be taken, which are—decide and take action, and measure for effectiveness. For all of the following techniques, the Safety Decision Hierarchy also applies.

RISK-ASSESSMENT MATRICES

For this primer, two-dimensional risk assessment matrices are discussed. They are displays of variations within two categories of terms: the *severity* of harm or damage that could result from a hazards-related incident, and the *probability* that the incident could occur.

In the chapter on risk-scoring systems that will soon follow, a review will be made of three- and four-dimensional risk assessment systems.

By agreeing on the meanings of the terms to be used in a risk assessment matrix, management establishes the levels of risk that require reduction and those that are acceptable. Also, risk assessment matrices can be used to compare and prioritize risks, and to effectively allocate mitigation resources.

Definitions of the levels of probability and severity used vary in actual practice because of the perceptions of those who develop the matrices; the inherent hazards of a product, process, or operation; and management's risk tolerance. Since the matrix is a management decision tool, management personnel at the appropriate level should agree on the related definitions.

Safety practitioners should understand that a fixed set of definitions of terms for incident probability and severity cannot be developed that meet the requirements of all situations. Thus, they should tailor a matrix that suits the hazards and risks and the management tolerance for risk with which they commonly deal. A variety of risk matrices in use are presented here to provide safety practitioners with a broad base of information from which choices can be made in developing the matrix considered appropriate for their needs.

The breadth of possibilities is extensive. Risk assessment matrices have been developed that apply to but one or a combination of several of the following injury or damage classes: employees, the public, facilities, equipment, product, operation downtime, and the environment.

To begin with, examples follow to illustrate the descriptions in use for the probability of occurrence and severity of consequence indicators. Tables 1 and 2

TABLE 1 Severity of Consequences. Adapted from MIL-STD-882D and Sverdrup Technology Publications

Category: Descriptive Word	People: Employees, Public	Facilities, Product, or Equipment Loss	Operation Downtime	Environmental Damage
Catastrophic	Fatality	Exceeds $1 M	Exceeds 4 months	Major event, requiring several years for recovery
Critical	Disabling injury or illness	250 K to $1 M	2 weeks to 4 months	Event requires 1 to 5 years for recovery
Marginal	Minor injury or illness	1 K to 250 K	1 day to 2 weeks	Recovery time is less than 1 year
Negligible	No injury or illness	Less than 1 K	Less than 1 day	Minor damage, easily repaired

TABLE 2 Incident Probability. Adapted from MIL-STD-882D and Sverdrup Technology Publications

Category: Descriptive Word	Definition: Applies for the Selected Unit of Time, Events, Population, Items, or Activity
Frequent	Likely to occur repeatedly
Probable	Likely to occur several times
Occasional	Likely to occur sometime
Remote	Not likely to occur
Improbable	So unlikely, can assume occurrence will not be experienced

are adapted from the "Standard Practice for System Safety" (MIL-STD-882D) and safety publications issued by Sverdrup Technology, Inc.

To repeat, obtaining agreement on what the terms mean and having them suit an industry or individual entity is important. Two examples are now presented of the definitions of probability and severity categories as they were given in the Risk Management Guide for the Aviation Industry in the *Aviation Ground Operation Safety Handbook* and the "Safety Guidelines for Risk Assessment" (SEMI S10-1296). Both guides apply to several injury and damage classes.

The terms used for probability and severity indicators in Tables 3 through 6 differ. And, in combination, they designate different risk levels, different remedial action levels, and different risk acceptance levels.

TABLE 3 Risk Management Guide for the Aviation Industry Severity Descriptions from Aviation Ground Operation Safety Handbook

Catastrophic	Death or permanent total disability, system loss, major property damage
Critical	Permanent, partial, or temporary disability in excess of 3 months, major system damage, significant property damage
Marginal	Minor injury, lost work-day accident, minor system damage, minor property damage
Negligible	First aid or minor medical treatment, minor system impairment

TABLE 4 Risk Management Guide for the Aviation Industry Probability Descriptions from Aviation Ground Operation Safety Handbook

Frequent	Occurs often, continuously experienced
Likely	Occurs several times, occurs often
Occasional	Occurs sporadically, occurs sometimes
Seldom	Remote chance of occurrence; unlikely but could occur at some time
Unlikely	Can assume it will not occur

TABLE 5 Safety Guidelines for Risk Assessment, SEMI S10-1296 Severity Grouping

Severity Group	People	Equipment/Facility	Property
Catastrophic	One or more fatalities	System or facility loss	Chemical release with acute long-lasting environmental or public-health impact
Severe	Disabling injury/illness	Major subsystem loss or facility damage	Chemical release with temporary environmental or public-health impact
Moderate	Medical treatment or restricted work activity (OSHA recordable)	Minor subsystem loss or facility damage	Chemical release triggering external reporting requirements
Minor	First aid only	Non-serious equipment or facility damage	Chemical release requiring only routine cleanup without reporting

TABLE 6 Safety Guidelines for Risk Assessment, SEMI S10-1296 Likelihood Grouping: Expected Rate of Occurrence

Frequent	More than five times per year
Likely	More than once per year, but no more than five times per year
Possible	More than once in five years, but no more than once per year
Rare	More than once in ten years, but no more than once in five years
Unlikely	No more than once in ten years

In the following three examples of risk assessment matrices, severity and probability descriptors are related so risk ratings can be made. First, an adaptation of the "mishap risk categories and mishap acceptance levels" in MIL-STD-882D will be made. This statement appears in the General Requirements section of the standard:

> When properly applied, these requirements should ensure the identification and understanding of all known hazards and the associated risks; and mishap risk is eliminated or reduced to an acceptable level.

In Appendix A of the standard, it is stated that the use of a mishap assessment matrix allows classification by mishap severity and mishap probability and assists in managing the decision making to achieve the necessary risk elimination or reduction to an acceptable level.

Table 7 varies somewhat in its style of presentation from the text of the standard (MIL-STD-882D), but not in substance. The next example of a risk assessment matrix appears in Table 8; it is based on "Safety Guidelines for Risk Assessment" (SEMI S10-1296). There are five risk assessment categories provided; in Table 7, there were four.

TABLE 7 Risk-Assessment Matrix from Standard Practice for System Safety, MIL-STD-882D

	Severity of Consequence			
Occurrence Probability	Catastrophic	Critical	Marginal	Negligible
Frequent	High	High	Serious	Medium
Probable	High	High	Serious	Medium
Occasional	High	Serious	Medium	Low
Remote	Serious	Medium	Medium	Low
Improbable	Medium	Medium	Medium	Low

TABLE 8 Risk-Assessment Matrix Based on SEMI S10-1296

	Likelihood				
Severity	Frequent	Likely	Possible	Rare	Unlikely
Catastrophic	Critical	Critical	High	Medium	Low
Severe	Critical	High	Medium	Low	Low
Moderate	High	Medium	Low	Low	Slight
Minor	Medium	Low	Low	Slight	Slight

If a matrix is developed containing these five risk assessment categories—critical, high, medium, low, slight—indicators of management decision levels should be established and agreed upon (as in Table 10). The third example of a risk assessment matrix is Table 9, adapted from "Risk Assessment and Reduction—A Guideline to estimate, evaluate and reduce risks associated with machine tools (B11.TR3). This document pertains principally to worker safety. In the abstract for B11.TR3, it is said that the risk reduction process is not complete until tolerable risk is achieved.

Remedial action or acceptance levels must be ascribed to the risk categories to permit intelligent decision making on the part of management. Possible management decision levels are illustrated in Table 10, to serve as the basis of reflection for those who craft a risk assessment matrix.

TABLE 9 Risk-Estimation Matrix Based on B11.TR3

	Severity of Harm			
Occurrence Probability	Catastrophic	Serious	Moderate	Minor
Very likely	High	High	High	Medium
Likely	High	High	Medium	Low
Unlikely	Medium	Medium	Low	Negligible
Remote	Low	Low	Negligible	Negligible

TABLE 10 Management Decision Levels

Risk Category	Remedial Action or Acceptance
High	Operation not permissible
Serious	Remedial action to have high priority
Medium	Remedial action to be taken in appropriate time
Low	Risk is acceptable: remedial action discretionary

Going through the exercise of creating and reaching agreement on a risk assessment matrix adds immensely to a safety practitioner's effectiveness in communicating about risks and obtaining serious consideration of the remedial actions being recommended. The same applies, no matter what hazard analysis and risk assessment technique is used.

DESCRIPTIONS OF HAZARDS ANALYSES AND RISK ASSESSMENT TECHNIQUES

In the previous chapter, reference was made to the hazards analyses methods mentioned in the following sources:

- The Risk Management Guide for the Aviation Industry in the Aviation Ground Operation Safety Handbook (there are six possible methods listed)
- "OSHA's Rule for Process Safety Management of Highly Hazardous Chemicals" (there are six provided, but they are not identical to those in the aviation industry guide)
- EPA's "Risk Management Programs for Chemical Accidental Release Prevention" (the analyses methods listed are identical to those in the OSHA standard)
- *Guidelines For Hazard Evaluation Procedures, Second Edition With Worked Examples* (there are 11 alternatives cited, 4 of which are included in the OSHA publication)

But, this list is not complete. As an indication of just how many methodologies have been developed, consider these two sources: Pat Clemens gives brief descriptions of 25 techniques in "A Compendium of Hazard Identification and Evaluation Techniques for System Safety Applications"; in the *System Safety Analysis Handbook*, 101 methods are described.

Brief descriptions will be given here of nine hazard analysis techniques. It is important to understand that each analytical technique discussed complements, rather than supplants, the others. Selecting a technique or a combination of techniques to analyze a hazardous situation requires good judgment based on knowledge and experience. Qualitative rather than quantitative judgments will prevail. And, for all but the complex risks, qualitative judgments will be sufficient. Sound quantitative data on event probabilities are seldom available.

Preliminary Hazard Analysis

The original purpose of the *preliminary hazard analysis* (PHA) technique was to serve as the initial effort in identifying and evaluating hazards in the early stages of the design process. However, in actual practice, the technique has attained broader use. The principles on which preliminary hazard analyses are based are used not only in the initial design process, but also in reassessing the safety of existing products or operations. For example, the hazard analysis and risk-assessment requirements of European standards EN 292 and EN 1050, which were discussed in the previous chapter, have been adequately met in some companies in the design or redesign stages by applying an adaptation of the preliminary hazard analysis technique.

Further, when simpler techniques (i.e., what-if analyses, checklist analyses) reveal hazardous situations requiring additional review, the substance of the preliminary hazard analysis technique may be applied. In reality, the technique needs a new name. (Also, take note: Within "OSHA's Rule for Process Safety Management of Highly Hazardous Chemicals" and EPA's "Risk Management Program for Chemical Accidental Release Prevention," PHA stands for *process hazard analysis*.)

Headings on preliminary hazard analysis forms will typically include identification data: date, names of evaluators, department, and location. The following information is also often listed:

- A hazard description, sometimes called a *hazard scenario*
- A description of the task, operation, system, or product being analyzed
- Which exposures are the subject of the analysis: people (employees, the public); facility, product, or equipment loss; operation downtime; environmental damage
- The probability interval to be considered: unit of time or activity; events; units produced; life cycle
- A numerical or alpha indicator for the severity of harm or damage that might result if the hazard's potential is realized
- A numerical or alpha indicator for the occurrence probability
- A risk-assessment code, using the agreed upon risk-assessment matrix
- Remedial action to be taken, if risk reduction is needed

A communication accompanies the analysis, explaining the assumptions made and the rationale for them. Comments would also be made on the assignment of responsibilities for the remedial actions necessary and a timeframe given. An example of a preliminary hazard analysis worksheet appears in Appendix A at the end of this chapter, courtesy of Sverdrup Technology, Inc. That form requires the entry of severity, probability, and risks codes before and after countermeasures are taken.

On Developing a Coding System

Assume that there are four categories of severity of consequences, for which numerical codes will be used:

Catastrophic: 1
Critical: 2
Marginal: 3
Negligible: 4

Assume also that there are five categories of occurrence probability, having alpha codes:

Frequent: A
Probable: B
Occasional: C
Remote: D
Improbable: E

The foregoing is an adaptation of the risk-assessment matrix shown earlier in Table 7. Table 11 is an extension of Table 7. Code indicators have been added to the matrix.

TABLE 11 Including Probability and Severity Codes in the Risk-Assessment Matrix. Adapted from Standard Practice for System Safety, MIL-STD-882D

	Severity of Consequence			
Occurrence Probability	Catastrophic: 1	Critical: 2	Marginal: 3	Negligible: 4
Frequent: A	High	High	Serious	Medium
Probable: B	High	High	Serious	Medium
Occasional: C	High	Serious	Medium	Low
Remote: D	Serious	Medium	Medium	Low
Improbable: E	Medium	Medium	Medium	Low

Risk codes would then be as follows, with various combinations of the severity and probability codes taken into account:

Combinations	Risk Category and Code
A-1, A-2, B-1, B-2, C-1	High: H
A-3, B-3, C-2, D-1	Serious: S
A-4, B-4, C-3, D-2, D-3, E-1, E-2, E-3	Medium: M
C-4, D-4, E-4	Low: L

A safety practitioner can make suitable adaptations from the above list.

Safety Reviews: Operations Analyses

Call them what you like—safety reviews, operations analyses, loss prevention reviews, or walk-through inspections—their purpose is to identify and analyze hazards. Sometimes, the probable consequences of hazardous events are also assessed. Hazards identification and analyses of this sort have historically been a part of the practice of a large percentage of safety practitioners. Individuals may make them informally in a brief time, or they can be completed by a team that spends weeks on one project. Involvement of location personnel may be extensive, depending on the review's purpose. Documentation may be limited or considerable.

The state of the art would advance if, more often, risk assessment and prioritization became a part of safety and operation reviews.

What-If Analysis

For a *what-if analysis*, a group of people (as little as two, but often several more) would use a brainstorming approach to identify hazards and hazard scenarios, how incidents can occur, and what their probable consequences might be.

Questions posed during the brainstorming session may commence with what-if, as in "What if the air conditioning fails in the computer room?" or may be expressions of more general concern, as in "I worry about the possibly of spillage and chemical contamination during truck offloading."

In a typical procedure, all the questions are recorded and assigned for investigation. Each subject of concern is then addressed by one or more team members. They would consider the potential of the hazardous situation and the adequacy or inadequacy of risk management controls in effect, suggesting additional risk reduction measures if appropriate.

Checklist Analysis

Checklists are primarily used when published standards, codes, and industry practices exist. There are many such checklists. They consist of lists of questions pertaining to the applicable standards and practices—usually with a yes, no, or not applicable response. Their purpose is to identify deviations from the expected and thus possible hazards. A *checklist analysis* requires a walk-through of the area to be surveyed.

Checklists are easy to use and provide a cost-effective way to identify customarily recognized hazards. Nevertheless, the quality of checklists is dependent on the experience of the people who develop them. Further, they must be crafted to suit particular needs. If a checklist is not complete, the analysis may not identify some hazardous situations. An example of a checklist for machinery design is provided in Appendix B at the end of this chapter. Another checklist, for general design purposes, appears in Chapter 7. They are good resources for those who wish to compile their own checklists.

What-If/Checklist Analysis

A *what-if/checklist hazard analysis* technique combines the creative, brainstorming aspects of the what-if method with the systematic approach of the checklist. The combination of techniques can compensate for the weaknesses of each. The what-if part of the process, involving brainstorming, can help the team identify hazards that have the potential to be the causal factors for incidents, even though no such incidents have yet occurred. The checklist provides a systematic approach for review that can serve as the generator of ideas during the brainstorming process. Usually, a team experienced in the design, operation, and maintenance of the operation performs the analysis. The number of people required depends, of course, on the operation's complexity.

Hazard and Operability Analysis

The *hazard and operability analysis* (HAZOP) technique was originally developed to identify both hazards and operability problems at chemical process plants. The technique, which has been in use for many years, requires an interdisciplinary team and an experienced team leader.

The purpose of a HAZOP is to review a process or operation systematically to identify whether deviations from the desired practices could lead to undesirable consequences. HAZOPs can be used at any stage in the life of a process.

HAZOPs usually require a series of meetings during which the team, using process drawings, systematically evaluates the impact of deviations from the desired practices. The team leader uses a set of guide words to develop discussions. As the team reviews each step in a process, they record any deviations, along with their causes, consequences, safeguards, and required actions, or the need for more information to evaluate the deviation.

Failure Modes and Effects Analysis

In several industries, *failure modes and effects analysis* (FMEA) has been the technique of choice by design engineers for reliability and safety considerations. They are used to evaluate the ways in which equipment fails and the response of the existing system to those failures. Although an FMEA is typically made early in the design process, the technique can also serve well as an analysis tool throughout the life of equipment or a process.

For each item of equipment, an FMEA produces qualitative, systematic lists that include the failure modes, the effects of each failure, safeguards that exist, and additional actions that may be necessary. For example, for a pump, the failure modes might include the following: fails to stop when required; stops when required to run; seal leaks or ruptures; and pump case leaks or ruptures. Both the immediate effects and the impact on other equipment would be recorded. Generally, when analyzing impacts, the probable worst case is assumed and analysts would conclude that existing safeguards do or do not work.

Although an FMEA can be completed by one person, it is typical for a team to be appointed when it will be a complex process. In either case, the traditional process is similar:

1. Identify the item or function to be analyzed.
2. Define the failure modes.
3. Record the failure causes.
4. Determine the failure effects.
5. Enter a severity code and probability code for each effect.
6. Enter a risk code.
7. Record the actions required to reduce the risk to an acceptable level.

Note that the FMEA process described here requires the entry of probability, severity, and risk codes. The risk-assessment matrix shown in Table 11 includes the codes necessary for traditional FMEA purposes. A typical FMEA form on which those codes would be entered is provided at the end of this chapter in Appendix C, courtesy of Sverdrup Technology, Inc.

The FMEA technique is principally a design and redesign analysis tool. Safety practitioners who become involved in the design aspects of safety would benefit greatly by having some knowledge of this technique.

To suit particular needs, the basic FMEA method has been extended to make it three-dimensional. For example, one extension adds criticality to the FMEA process; another adds a detection criterion. Both will be discussed in the next chapter.

Fault Tree Analysis

A *fault tree analysis* (FTA) is a top-down, deductive logic model that traces the failure pathways for a predetermined, undesirable condition or event, called the *TOP event*. An FTA can be carried out either quantitatively or subjectively.

The FTA generates a fault tree (a symbolic logic model) entering failure probabilities for the combinations of equipment failures and human errors that can result in an accident. Each immediate causal factor is examined to determine its subordinate causal factors until the root causal factors are identified. The strength of an FTA is its ability to identify combinations of basic equipment and human failures that can lead to an accident, allowing the analyst to focus preventive measures on significant basic causes.

An FTA has particular value when analyzing highly redundant systems and high-energy systems in which high-severity events can occur. For systems vulnerable to single failures that can lead to accidents, the FMEA and HAZOP techniques are better suited. FTA is often used when another technique has identified a hazardous situation that requires more detailed analysis. Making a fault tree analysis of other than the simplest systems requires experienced analysts.

Management Oversight and Risk Tree

As Clemens wrote in a previously cited paper (15), the *management oversight and risk tree* (MORT) technique applies "a pre-designed, systematized logic tree to the identification of total system risks, both those inherent in physical equipment and processes and those which arise from operational/management inadequacies." MORT is an incident investigation and analysis technique. It is discussed here for a particular purpose. There are three major elements in the practice of safety:

- In the design processes: preoperational
- In the operational mode: within a continuous improvement process
- Post incident: through investigation of hazards-related incidents for causal factor determination

All the hazard analysis and risk assessment techniques previously discussed relate principally to the initial design process or the redesign process for achieving risk reduction in the operational mode *before hazards-related incidents occur*. MORT was developed principally for the purpose of *incident investigation*. In the Abstract for the *Guide to Use of the Management Oversight and Risk Tree*, this is how MORT is described:

> MORT is a comprehensive analytical procedure that provides a disciplined method for determining the systemic causes and contributing factors of accidents. MORT directs the user to the hazards and risks deriving from both system design and procedural shortcomings.

This is an excerpt from the "Performance Errors" section (p. 19), also in the *Guide*:

> It should be pointed out that the kinds of questions raised by MORT are directed at systemic and procedural problems. The experience, to date, shows there are few "unsafe acts" in the sense of blameful work level employee failures. Assignment of "unsafe act" responsibility to a work level employee should not be made unless or until the preventive steps of: (1) hazard analysis, (2) management or supervisory direction, and (3) procedures for safety review have been shown to be adequate.

MORT provides an excellent hazard analysis method for the postincident element of the practice of safety. Knowing how to use the method contributes to the completion of an incident investigation, which is an important aspect of safety. Unfortunately, research shows that the quality of typical incident investigation is too often deficient. (See the chapter entitled, "Incident Investigation: Studies of Quality," in *On the Practice of Safety*.)

SOFTWARE

There are several software products available that assist in making hazards analyses and risk assessments. Comments follow on a select few of them.

designsafe: the hazard analysis and risk assessment guide. The Hazard Analysis and Risk Assessment Guide. A general application software program that supports several of the hazard analysis and risk assessment techniques discussed in this chapter. The first part walks safety practitioners and design engineers through a hazard analysis, helping to identify the hazards associated with the area or product selected for analysis. The second part, a risk assessment, helps users analyze the hazards previously identified. The program helps prioritize hazards according to the analyst's input on probability of occurrence and severity of injury or damage.

designsafe also helps ensure that hazards are fully addressed by explicitly identifying hazards that have not been resolved. Several reports can result via this program: a risk level report, a status report, and a hazard analysis report.

This software program has been successfully used in a broad range of applications: e.g., the design, installation, use, repair, and retirement of systems, equipment, facilities, and products. At www.designsafe.com, both additional information and a free demo are available.

"Design for Safety" Toolbox. Assists in designing facilities with the construction worker's health and safety as the foremost considerations. It is available through the Construction Industries Institute, whose Internet address is www.construction-institute.org.

SEMATECH Failure Modes and Effects Analysis (FMEA) Software Tool. Provides just what the title indicates, a software tool to assist in making an FMEA. At www.sematech.org, look for Technology Transfer #92091302A-XFR.

FaultrEASE. A graphic tool for creating, editing, and computing fault trees. A colleague said a remarkable thing about the manual: It is readable. The program makes layout and pruning the tree relatively easy, for those skilled in fault tree analysis. It is available in both Macintosh and Windows. Visit www.process-safety.com.

AVOIDING UNREALISTIC EXPECTATIONS

Making hazards analyses and risk assessments is both art and science. Whatever the methodology—the simplest or the most complicated—much judgment will be used in determining the severity potentials of hazards and the probability of incidents occurring. Even though those applying the methodologies make learned judgments, they may disagree on which hazards are most important because of their severity potential and which risks deserve the highest priority.

One way to resolve these differences is to have qualified teams participate when the hazards and risks are considered significant, the intent being to reach a consensus.

Some who oppose the use of these methodologies do so because the outcomes are not stated in absolutely precise numbers. Such accuracy is not attainable, and expecting such results is unrealistic. Fortunately, recognition continues to grow that hazard analysis and risk assessment methods add considerable value to safety decision making.

CONCLUSION

This chapter is but a primer on hazard analysis and risk assessment. Its purpose is to provide a foundation for those who perceive that additional knowledge in this aspect of safety provides an opportunity for professional growth, accomplishment, and recognition. Having such knowledge adds to one's ability to evaluate hazardous situations and make more convincing presentations to management for resource allocation in accomplishing the risk reduction measures proposed.

It is also realistic to suggest, in relation to the growing interest in hazard analysis and risk assessment, that safety practitioners become proficient in preliminary hazard analysis, in addition to the more commonly used techniques—safety reviews, what-if analysis, checklist analysis. Having an understanding of failure modes and effects analysis would also be beneficial. For example, when making an ergonomics analysis, the FMEA methodology can be applied, except that the effects deriving from design and operational shortcomings are parts of the human body.

REFERENCES

Aviation Ground Operation Safety Handbook, 5th ed. Itasca, IL: National Safety Council, 2000.

Clemens, Pat. "A Compendium of Hazard Identification and Evaluation Techniques for System Safety Applications." *Hazard Prevention*, March/April 1982.

Failure Mode and Effects Analysis (FMEA): A Guide for Continuous Improvement for the Semiconductor Equipment Industry. SEMATECH. Austin, TX, 1992. See http://www.sematech.org.

Framework for Environmental Health Risk Management, Final Rept., Vol. 1. Washington, DC: The Presidential/Congressional Commission on Risk Assessment and Risk Management, 1997.

Guide to use of the Management Oversight and Risk Tree, SSDC-103. Washington, DC: Dept. of Energy, 1994.

Guidelines for Hazard Evaluation Procedures, Second Edition with Worked Examples. New York: Center for Chemical Process Safety of the American Institute of Chemical Engineers, 1992.

Johnson, William G. *MORT Safety Assurance Systems.* New York: Marcel Dekker, 1980.

Manuele, Fred A. *On the Practice of Safety*, 2nd ed. New York: John Wiley & Sons, 1997.

"OSHA's Rule for Process Safety Management of Highly Hazardous Chemicals." 1910.119 of 29 CFR 1910, Washington, DC: Dept. of Labor, OSHA, 1992.

Risk assessment and reduction—a guideline to estimate, evaluate and reduce risks associated with machine tools. McLean, VA: The Association for Manufacturing Technology, 2000.

"Safety of Machinery—Basic Concepts, General Principles for Design; Part 1. Basic Terminology, Methodology." European Standard EN 292-1, Brussels: Central Secretariat, European Committee for Standardization, 1991.

"Safety of Machinery—Basic Concepts, General Principles for Design; Part 2. Technical Principles and Specifications." European Standard EN 292-2, Brussels: Central Secretariat, European Committee for Standardization, 1991 (amended 1997).

"Safety of Machinery—Principles for Risk Assessment." European Standard EN 1050, Brussels: Central Secretariat, European Committee for Standardization, 1997.

"Safety Guidelines for Risk Assessment." SEMI S10-1296, Mountain View, CA: Semiconductor Equipment and Materials International, 1996.

"Standard Practice for System Safety." MIL-STD-882D, Washington, DC: Dept. of Defense, 2000.

Stephans, R. A. and W. W. Talso. *System Safety Analysis Handbook*, 2nd ed. Albuquerque, NM: New Mexico Chapter of the System Safety Society, 1997.

Sverdrup Technology, Inc. Safety literature at http://www.sverdrup.com/safety.

ADDITIONAL READING

Grose, Vernon L. *Managing Risk, Systematic Loss Prevention for Executives.* Englewood Cliffs, NJ: Prentice-Hall, 1987 (now available through Omega Systems Group, Incorporated, Arlington, VA).

Raheja, D. G. *Assurance Technologies—Principles and Practices.* Englewood Cliffs, NJ: Prentice-Hall, 1991.

Roland, Harold E. and Brian Moriarty. *System Safety Engineering and Management*, 2nd ed. New York: John Wiley & Sons, 1990.

Stephenson, Joe. *System Safety 2000.* New York: Van Nostrand Reinhold, 1991.

Appendix A

Preliminary Hazard Analysis Worksheet

APPENDIX A

Sverdrup Technology, Inc. **Preliminary Hazard Analysis**

Brief Descriptive Title (Position of System/Sub-system/Operational Phases covered by this analysis):

Probability Interval: 25 years	Date:
System Number: ________	Analysis: ☐ Initial ☐ Revision ☐ Addition

Hazard No./Description	Hazard Target*	Risk Before: Severity	Risk Before: Probability	Risk Before: Risk Code	Description of Countermeasures Identify countermeasures by appropriate code letter(s): D = Design Alteration E = Engineered Safety Feature S = Safety Device W = Warning Device P = Procedures/Training	Risk After: Severity	Risk After: Probability	Risk After: Risk Code

Prepared by/Date:	*Target Codes: P—Personnel E—Equipment T—Downtime R—Product V—Environment	Approved by/Date:

Appendix B

Examples of Hazards, Hazardous Situations, and Hazardous Events

This checklist is an adaptation from Annex A in the European Standard EN 1050: Safety of machinery—Principles of risk assessment. The intent of the checklist is to serve as a guide for those who design and manufacture equipment and machinery that goes into European work places. Although it pertains to a broad range of equipment for all sorts of employment, those who use it as a reference, must understand that it could not possibly include all hazards or all hazardous situations.

Mechanical hazards due to machine parts or workpieces, e.g.,

- Shape
- Relative location
- Mass and stability (potential energy of elements which move under the effect of gravity)
- Mass and velocity (kinetic energy of elements in controlled or uncontrolled motion)
- Inadequacy of mechanical strength—accumulation of energy inside the machinery, e.g.,
  - elastic elements (springs)
  - liquids and gases under pressure
  - the effect of vacuum

Mechanical hazards due to the potential for

- Crushing
- Shearing

- Cutting or severing
- Entanglement
- Drawing-in or trapping
- Impact

Electrical hazards due to

- Contact of persons with live parts (direct contact)
- Contact of persons with parts which have become live under faulty conditions (indirect contact)
- Approach to live parts under high voltage
- Electrostatic phenomena
- Thermal radiation or other phenomena such as the projection of molten particles and chemical effects from short circuits, overloads, etc.

Thermal hazards, resulting in

- Burns, scalds and other injuries by a possible contact of persons with objects or materials with an extreme high or low temperature, by flames or explosions and also by the radiation of heat sources
- Damage to health by hot or cold working environment

Hazards generated by noise, resulting in

- Hearing loss (deafness), other physiological disorders (e.g., loss of balance, loss of awareness)
- Interference with speech communication, acoustic signals, etc.

Hazards generated by vibration

- Use of hand-held machines resulting in a variety of neurological and vascular disorders
- Whole body vibration, particularly when combined with poor postures

Hazards generated by radiation

- Low frequency, radio frequency radiation, micro waves
- Infrared, visible and ultraviolet light
- X and gamma rays
- Alpha, beta rays, electron or ion beams, neutrons
- Lasers

Hazards generated by materials and substances (and their constituent elements) processed or used by the machinery

- Hazards from contact with or inhalation of harmful fluids, gases, mists, fumes, and dusts
- Fire or explosion hazards
- Biological or microbiological (viral or bacterial) hazards

Hazards generated by neglecting ergonomic principles in machinery design

- Unhealthy postures or excessive effort
- Hazardous situations due to lifting
- Inadequate consideration of hand-arm or foot-leg anatomy
- Inadequate local lighting
- Mental overload and underload, stress
- Human error, human behavior
- Inadequate design, location or identification or manual controls
- Inadequate design or location of visual display units

Hazards deriving from unexpected start-up, unexpected overrun/overspeed (or any similar malfunction) from

- Failure/disorder of the control system
- Restoration of energy supply after an interruption
- External influences on electrical equipment
- Other external influences (gravity, wind, etc.)
- Errors in the software
- Errors made by the operator (due to mismatch of machinery with human characteristics and abilities)
- Impossibility of stopping the machine in the best possible conditions
- Variations in the rotational speed of tools
- Failure of the power supply
- Failure of the control circuit
- Errors of fitting

Hazards linked to the work position (including driving station) on the machine

- Fall of persons during access to (or at/from) the work position
- Exhaust gases/lack of oxygen at the work position
- Fire (flammability of the cab, lack of extinguishing means)

- Mechanical hazards at the work position
  - contact with the wheels
  - rollover, lack of stability
  - fall of objects, penetration of objects
  - break-up of parts rotating at high speed
  - contact of persons with machine parts or tools
- Insufficient visibility from the work position
- Inadequate lighting
- Inadequate seating
- Noise at the work position
- Vibration at the work position
- Insufficient means for evacuation/emergency exit

Hazards resulting in load falls, collisions, or machine tipping caused by

- Lack of stability
- Uncontrolled loading—overloading—overturning limits exceeded
- Uncontrolled amplitude of movement
- Unexpected/unintended movement of loads
- Inadequate holding devices/accessories
- Collision of equipment or machines
- Derailment
- Insufficient mechanical strength of parts
- Inadequate design of pulleys and drums
- Inadequate selection of chains, ropes, and accessories and their inadequate integration into the equipment or machine
- Lowering the load under friction brake
- Abnormal conditions of assembly/testing/use/maintenance
- Effect of a load on persons—impact of the load or counterweight

Additional hazards—including the driving function

- Break-up during operations
- Falling or ejected objects or fluids
- Loss of stability/overturning of machinery
- Movement when starting the engine
- Movement without a driver at the driving position
- Unauthorized start-up/use
- Movement without all parts in a safe position
- Excessive speed of pedestrian controlled machinery

- Excessive oscillations when moving
- Insufficient ability of machinery to be slowed down, stopped and immobilized
- Lack of or inadequacy of visual or acoustic warning systems
- Hazards from coupling and towing
- Hazards from the engine and batteries
- Insufficient instructions for the driver/operator

Appendix C

Failure Modes & Effects Analysis Worksheet

Source: Sverdrup Technology, Inc., with permission.

APPENDIX C

Project No.: ________
Subsystem: ________
System: ________
Probability Interval: ________
Operational Phase(s): ________

Sverdrup Technology, Inc.
Failure Modes & Effects Analysis

FMEA No.: ________

Sheet ____ of ____
Date: ________
Prep. By: ________
Rev. By: ________
Approved By: ________

IDENT. NO.	ITEM/ FUNCTIONAL IDENT.	FAILURE MODE	FAILURE CAUSE	FAILURE EFFECT	TARGET	RISK ASSESSMENT			ACTION REQUIRED/REMARKS
						SEV	PROB	RISK CODE	

Sverdrup Technology, Inc. Worksheet

P: Personnel/E: Equipment/T: Downtime/R: Product/V: Environment

7

How to Avoid Bringing Hazards into the Workplace

INTRODUCTION

When safety practitioners buy into the idea that the most effective and economical way to minimize risks is to have the hazards from which they derive addressed in the design process, they often logically ask: How can I become influential and prevent hazards from being brought into the workplace?

Within a few organizations, safety practitioners are influential because extensive procedures are in place in which they have a specified responsibility for capital expenditure proposal reviews, project design reviews, writing purchasing specifications for new equipment, and signing off on new or altered equipment before it can be placed in operation. In other entities, such procedures are not documented but followed, because safety practitioners have been successful in convincing engineering personnel that their counsel can provide value. In either case, safety professionals developed an expertise from which they could demonstrate meaningful participation and value. More often, safety practitioners are not involved in the design of facilities, equipment, or processes. For them, the opportunities to convince others of their worth by initiating successful participation are great.

The methods used to avoid bringing hazards into the workplace will be broadly explored here, with the hope that the materials covered will give safety practitioners relevant ideas which they can then adapt to their advantage. We note that there is no universally adopted set of guidelines on how to achieve participation in the processes that result in hazards not being brought into the workplace, and that those who have been successful as participants in those processes sometimes achieved such results by good planning, a careful assessment of their environment, and their own imagination and tactful aggressiveness. This chapter will include:

- Comments on how some safety practitioners are engaged in activities that lessen the probability of hazards being brought into the workplace
- Review of the procedures the auto industry and UAW have agreed to establish for hazard avoidance in the design process—for its significance and broad influence
- Edited composite of procedures in place in several companies to achieve hazard avoidance and control in the design process, and procedures to be followed before modified equipment is released for operation
- General checklist to serve as a reference by which a specifically tailored checklist can be developed for use in initial design reviews and equipment acceptance

HOW SOME SAFETY PRACTITIONERS ARE ENGAGED IN HAZARD AVOIDANCE

To obtain current information on how safety practitioners are involved in activities to avoid bringing hazards into the workplace, the author solicited comments on the subject through an Internet safety server. Here are some of the responses—the most unfortunate first:

- From an industrial hygienist: If the engineers would only let me into the design process, I know that many of the health hazards I deal with could be better controlled, we wouldn't need to do so much testing, and our operations would be more productive because we would reduce the amount of time employees spend on testing and on personal protective equipment usage.
- We don't have any forms or established procedures for our getting into what engineers are designing, and there is no formal method for engineering to notify us of a project. We do get copies of the engineering weekly reports, and we read those carefully. Then, we invite ourselves into the discussion.
- There was a lot of resistance by engineering to our getting into what they were designing because they had not recognized that we could be an asset. It took us a long time and quite a few money-saving successes to get it ingrained into their procedures that it was to their benefit to refer capital expenditure requests to us for our input. We don't have a written procedure, but it gets done.
- Almost all of our engineering is done by outside firms. I haven't been successful in convincing the few engineers we have left that I can help them write specifications that will save them money. But I have convinced the

manager of my plant that I am to have sign-off authority for new installations. My sign-off reviews can be embarrassing to the people who laid out the specs and I have to be very diplomatic in how I do what I need to do. But the plant manager gives me support now. He's being educated.

- From a construction safety professional: I request to see all drawings at the 10% level while changes are still easily incorporated. I get a schedule of all construction plans and visit contractor lay-down areas to inspect materials being used on the job. I attempt to educate quality assurance personnel on what to look for as safety indicators. I visit the engineering department and the contracting department on a daily basis when I'm in town. Yes, it is time-consuming but it saves money and lives as well as equipment. It took maximum effort on my part to get where I am.
- In my company, it's in the capital expenditure procedure manual that all funding requests will receive a safety review. Managers have the same responsibility for safety as they do for productivity and quality, and when they approve the capital expenditure request, they are also signing off for safety. But, my name appears on the capital expenditure distribution list among the management people who have to sign off as approving the request. This isn't as burdensome as it was in the beginning because I have educated a fairly stable engineering and management staff on what it takes to get my approval.
- Through our successes in ergonomics, engineering recognized our contributions not only for safety, but also for productivity. They now invite us into the design process in the idea stage. They make it plain that they look to us to see that they don't mess up. We learned the hard way over a lot of years that it is expensive to correct hazards that are in the machines and equipment we buy. We had to do some costly equipment modifications for employee safety and health, and for environmental situations after the installations were completed and in operation. Our engineers aren't easy on us. But that's okay. We had to accept the criticism from them that our ergonomics checklists were so general that they weren't helpful. I'm sending you the revised forms for new machine and equipment purchases, forklifts, and hand tools.
- Our Facility Safety Manual includes extensive procedures for documented reviews for safety, health, environmental, and ergonomic standards before budget approval is obtained on a new project, and for equipment reviews before the equipment is released to normal production. As you will see from what I'm mailing to you, we are deep into specifications, and the signature of a safety specialist is required in the project and budget review procedures.

This list shows that the involvement of safety practitioners in the specification writing, design, purchasing, and sign-off processes varies from nothing, to their being required to do so by written procedures. In all but one of the cases cited, there is some involvement by safety practitioners. And, that came about because they took the initiative and proved their value.

For those who try to make their presence felt in terms of avoiding the introduction of hazards in the workplace, consider the following as a starting point from which to prove your value: It is the author's experience that ergonomics presents a fertile field in which to demonstrate competence that results not only in risk reduction, but also in the achievement of improved productivity and lower costs.

AUTO INDUSTRY/UNITED AUTO WORKERS CONCEPTS AND THEIR IMPLICATIONS

Although the following excerpt is from the 1999 General Motors (GM)/United Auto Workers (UAW) labor agreement, similar wordings appeared in previous contracts with GM and other automakers to accomplish similar purposes. Why cite this agreement? Because it addresses how to avoid bringing hazards into the workplace. The agreed upon concepts begin with the design process and extend to the involvement of safety practitioners at the plant level.

> ... As early as possible and preferably in the zero phase of the planning in the design process ... the parties agree to perform Task Based Risk Assessments on new equipment and manufacturing systems, and on existing equipment and manufacturing systems where locally agreed to and approved by the Plant Safety Review Board. A Task Based Risk Assessment will be performed after the detailed designs are completed.... A review of anticipated equipment and/or processes with the shop committee and the Local Joint Health and Safety Committee will be held.
>
> The local Joint Health and Safety Committee may be required to travel to vendors, plants, or other locations to participate in a Design Review of such equipment or processes as outlined in the Design for Health and Safety Specification.
>
> Machinery, equipment or processes will not be released for production without the written approval of the Plant Safety Administrator.

In summary, this agreement presents a theoretical ideal, at least on paper. It indicates the following:

- Design specifications for safety and health are in place.
- Risk assessments are to be made as early as possible in the design phase.
- Risk assessments are to be made on existing equipment where locally agreed to.
- Safety and health practitioners may visit vendor locations for design reviews.
- Operation of machinery, equipment, or processes requires a written approval for release into production.

These provisions impact on safety at GM locations, on the standards to be met by firms engaged by GM to perform design and engineering work, and on the vendors of equipment supplied to GM. They also expand the knowledge and skill requirements of safety practitioners.

Each of the automakers with which the UAW has a contract has established procedures to accomplish what the contract cited here requires. At GM, one of the steps taken was to issue a specifications document that sets forth the requirements for the design and redesign of equipment and systems to achieve a safe operating environment. This document requires the following:

- Task-based risk assessments be made in the design process when new programs are initiated
- A safety buy-off be performed before a builder or fabricator of equipment or systems ships equipment
- A multidiscipline validation team assure that the desired level of safety has been met when the equipment or process is installed, using safety checklists and the task-based risk assessment summary—all to provide another look at the production system to verify that nothing has been missed
- Task-based risk assessments be made of existing equipment, as directed by the Plant Safety Review Board

It is not suggested that by following these directives, a system will work perfectly. But, if what is written in the auto industry/UAW contracts became the norm throughout all of American business, a giant step forward will have been taken for worker safety. Safety practitioners looking for additional ideas may find them in these agreements and the procedures established to implement them.

Although the reference here pertains to a union-employer contract, the methods defined are universally applicable and compatible with the emerging emphasis on safety through design.

A GENERAL INDUSTRY GUIDE: SAFETY DESIGN AND OPERATIONAL REQUIREMENTS

A composite is provided here of the procedures in place at three companies for design reviews and for a safety sign-off before new or modified equipment can be released into normal operations. Those procedures serve to avoid bringing hazards into the workplace. In a way, these requirements represent culture statements: Management has decided that hazards and risks must be addressed as equipment is designed and before it can be placed in operation.

This composite presents a basis for thought as safety practitioners pursue similar concepts and procedures. For additional reference, we include a preliminary safety review form (Figure 2), an equipment acceptance safety review form (Figure 3), a vendor/design engineering review form (Figure 4), and a general design safety checklist (Figure 5).

Requirements of Equipment And Process Design Safety Reviews.

A. Purpose

To establish procedures to ensure that hazards are analyzed and that risks are at an acceptable level when considering new, redesigned, and relocated equipment or processes.

B. Scope

These procedures apply to all equipment and processes that may present risks of injury to people, or damage to property or the environment. They pertain to the design or redesign of all new, transferred, and relocated equipment and processes.

For all aspects of these procedures, documentation shall be appropriate to the activity.

Safety will be used in these procedures as a term to encompass risks of injury or damage to personnel (employees and the public), property, and the environment.

C. Responsibilities

The Location Manager. The ultimate responsibility for safety rests with the location manager.

Project Manager. The project manager is responsible for assuring the following:

- Corporate safety requirements are met.
- Safety documentation to accompany capital expenditure requests is prepared.
- Preliminary and subsequent design safety reviews are conducted.
- Appropriate coordination and communication take place with outside design and engineering firms to assure that specifications are met.
- Proper consideration is given to any safety problems identified by the staff during their visits to vendors and design and engineering firms prior to delivery of equipment.

Design Engineers. Principle responsibility of the design engineers is to design inherently safer equipment and processes. Whether employees or contractors, design engineers shall assure that the factors necessary for safety have been considered during the design process. They will provide the project manager and safety review team documentation, including:

- Detailed equipment design drawings
- Equipment installation, operation, preventive maintenance, and test instructions
- Details of and documentation for codes and design specifications

- Requirements and information needed to establish regulatory permitting and/or registrations

Safety Review Team. This team will conduct preliminary and subsequent design safety reviews for equipment and processes, or have design reviews made by outside consultants for particular needs.

In addition to the project manager, members will include the project design engineers, production and maintenance personnel, the facilities engineer, selected disinterested engineers, the safety professional, and others (financial, purchasing) as needed.

The safety review team will also be responsible for:

- Arranging and conducting safety walk-downs of new projects
- Determining when visits are to be made at vendor design and engineering locations, and selecting personnel with the necessary skills to make the visits

Safety Professional. The safety professional will:

- Serve as a member of the safety review team and assist in identifying and evaluating hazards in the design process and provide counsel as to their avoidance, elimination, or control
- Visit design engineering firms and other vendors, when requested by the safety review team, to assure that any safety problems are identified and corrected prior to shipment of equipment
- Sign off on an equipment acceptance safety review form* prior to newly installed or altered equipment or processes being released to normal production

Department Managers and Supervisory Personnel. Department managers and supervisors will give support to the project manager for the activities that come under their jurisdiction. They will be signatories, along with engineering personnel and the safety professional, on sign-off forms before newly installed or altered equipment is released for normal operation.

D. Initial Capital Expenditure Safety Review

Capital expenditure requests at financial levels requiring divisional or corporate approval are to be accompanied by a preliminary safety review form**, which is to be completed by the safety review team. Comments would be included in the form, assuring that the hazards and risks identified would be properly addressed.

For new or altered equipment or processes at a financial level not requiring

* See Figure 2 in text as an example.

** See Figure 1 in text as an example.

divisional or corporate approval, formal preliminary safety reviews are to be made at the discretion of the safety review team.

E. Design Reviews

When designs have been completed and drawings and specifications are available, the safety review team will hold one or more design review meetings to:

- Identify hazards not given appropriate attention, and recommend solutions to attain acceptable risk
- Assure that corporate safety requirements are being met
- Avoid the cost of risk reduction retrofitting as the project moves forward
- Assure compliance with applicable regulations, codes, and standards.

F. Safety Walk-Downs

When a project reaches completion at approximately a 70% level, the safety review team will arrange and conduct a safety walk-down to provide an opportunity to:

- Assure that specifications have been met
- Determine that hazards identified in the preliminary safety review and the subsequent design review have been properly addressed
- Identify hazards that may have been built inadvertently into the project, and to arrange for the necessary action to be taken

Similarly, a final safety walk-down will be arranged as the project nears completion.

H. Equipment and Process Release Requirements

For newly purchased, redesigned, or relocated equipment or processes—before release for normal production:

- Task analyses shall be made to identify hazards and risks, and the hazards identified are to be properly dealt with.
- Any revisions necessary in the written job procedure shall be made.
- Retraining as required shall be given.
- An equipment acceptance safety review form shall be completed; the signatories of this shall be the engineer-in-charge, the department manager or supervisor, and the safety professional.

I. Design Reviews at Vendor and Design Engineering Locations

When considered advantageous by the safety review team, arrangements will be made for personnel with the appropriate skills to visit vendors and design engineering firms to:

- Assure that vendors are building equipment to specifications
- Determine whether hazards exist that were not identified in the design review process, or by the vendor or design engineering firm, which need attention
- Particularly, avoid the high cost of retrofitting for safety matters during installation, testing, and debugging

For these reviews, sections of the general design safety checklist appropriate to the task will be downloaded to create a specifically tailored vendor/design engineering review form. Reports prepared as a result of such visits shall be made available to the safety review team and the project manager.

Preliminary Safety Review Form

A preliminary safety review goes by several names. In one company, its purpose is to meet "fitness for use criteria." In another company, it is referred to as a "hazards screening analysis." A *preliminary safety review form* is a listing of subjects pertaining to equipment, facilities, or processes that aids the reviewers in identifying hazards and risks which must be addressed. See Figure 1. Completion of the form produces, in effect, an early design review. The review team indicates whether a subject needs further consideration. No one list is suitable for all needs. In drafting such a list, a safety professional should use the general design safety checklist as a reference and include some or all of the topics presented in Figure 1, perhaps even additional ones.

Ergonomics	Material handling
Machine guarding	Illumination needs
Fire protection	Means of egress
Walking and working surfaces	Confined spaces
Use of hoists, cranes, etc.	Work at heights
Electrical potentials	Lockout/tagout
Confined spaces	Temperature extremes
Hazardous or toxic materials	Noise or vibration
Personal protective equipment	Nonionization emitters
Environmental concerns	Sanitation

FIGURE 1 *Preliminary safety review form.*

Equipment Acceptance Safety Review Form

Drafting acceptance safety review forms that are specific to every piece of equipment or process would be a mammoth undertaking. Nevertheless, industry-specific safety review checklists do exist and they should be used where applicable. The example of an *equipment acceptance safety review form* shown

Dept.________________ Control No.________________

Equipment Description____________________________

This form must be completed prior to equipment being released to normal production. It is applicable to:

1. All newly installed equipment or processes
2. Changes made in the use of equipment or processes
3. Modifications of existing equipment or processes

Preliminary approval indicates that the equipment is ready for initial production trials, but needs additional work for safety as listed in a memo attachment entitled, "Safety Items Needing Attention." Final approval indicates that the preliminary findings have been addressed satisfactorily and the equipment can be released to normal production.

	Preliminary	**Final**
Signed: ____________ Engineer-in-charge	______ Date	______ Date
Signed: ____________ Dept. Mgr. or Supvisor.	______ Date	______ Date
Signed: ____________ Safety Manager	______ Date	______ Date

The original copy of this form shall be retained by the Department Manager or Supervisor. The Engineer-in-charge and the Safety Manager will retain copies. The company's file retention policy shall apply.

Form 22 10/01 1,000

FIGURE 2 *Equipment acceptance safety review form.*

in Figure 2 assumes the existence of a general design safety checklist that can serve as a reference for the review process. Also, the example presumes two levels of review: a preliminary review to identify items needing attention and a final sign-off.

Design Safety Reviews

In Chapter 13 entitled, "Application in General Industry," of *Safety Through Design*, Paul Adams does a good job of commenting on design safety reviews. Excerpts appear, with permission (160).

Formal design safety reviews may seem tedious; however, they are critical for the delivery of inherently safe designs. The participants are all stakeholders and share responsibility for the safety of the final design. Most experienced engineers approach design reviews as an important problem-solving event. A spirit of cooperation, and even fun, is maintained by restricting criticism to constructive debate on specific design features.

Although the focus is always on safety, review teams frequently identify additional opportunities to reduce costs and improve productivity. Constructibility and maintainability issues often surface, and these alone can pay huge dividends for the comparatively short time invested in reviewing a project. These sessions also provide some of the best possible training for both novice and experienced engineers, as collective knowledge and experience are openly shared.

The standard approach for conducting a formal design safety review is to methodically work through a General Design Safety Checklist, and additional checklists for specific disciplines as necessary. For each system element, reviewers address the various forms of energy present and the steps taken to control unwanted release. A typical design safety review meeting proceeds as follows:

1. Project manager distributes drawings and copies of checklists.
2. Review team chooses a facilitator, often a disinterested engineer.
3. Project engineer/manager describes the project and its scope, and answers general questions about major areas of concern.
4. Project engineer/manager keeps notes.
5. Facilitator leads a methodical review using a generic checklist, with team members asking detailed questions to ensure thorough consideration of hazards and their control. Checklist items and sections that are not applicable are so noted.
6. Additional discipline checklists are reviewed as appropriate.
7. A marked-up copy of the checklists is retained with project documents.

Specific deliverables from a design review include:

- A set of marked-up drawings and specifications.
- A list of design modifications requiring attention prior to release for bid.
- A list of specific items to be checked during installation and the final walk-down. (There are often items identified as potentially problematic that cannot be adequately assessed at the time of the review due to lack of detailed knowledge.)
- Assignments for resolving specific details or making design corrections.

Following the review, it is the responsibility of the project engineer/manager to follow up and ensure that all issues raised during the design safety review are resolved and appropriate revisions completed.

This is what a design safety review is all about. Done well, such reviews serve brilliantly to minimize bringing hazards into the workplace. One important point: Choosing the right facilitator is crucial. In some companies, it is standard practice to engage a consultant to lead safety design reviews to assure that one individual's views do not dominate, and that all participants are given an opportunity to be heard.

General Design Safety Checklist

A detailed, specifically referenced design checklist covering all workplace safety needs would fill thousands of pages. The *general design safety checklist* presented here is a brief composite taken from several sources. Some safety professionals will view it as excessive; others will find that it does not address all their needs. Those who use it as a reference should be aware that there are many subject-specific and industry-specific checklists to which they can also refer.

Two such examples are given here: *Guidelines for Hazard Evaluation Procedures*, 2nd edition, issued by the Center for Chemical Process Safety, that includes a checklist questionnaire for chemical operations which fills 45 pages; and, the checklist provided in Appendix B at the conclusion of Chapter 9, which is adapted from European Standard EN 1050, "Safety of Machinery—Principles of Risk Assessment." It serves as a guide for those who design and manufacture equipment and machinery that goes into European workplaces.

The general design safety checklist is intentionally presented as a list of questions without the boxes and lines typically found in checklists. Also, all response columns that would generally fall to the right of a design checklist form have been eliminated in which users would enter check marks for—yes, no, or not applicable.

General Design Safety Checklist

Preface

This checklist begins with a preface that brings attention to Haddon's unwanted energy release theory, which modified and extended for workplace use, is this: For all injuries or illnesses, the transfer of energy to the human body, or exposure to a harmful environment, is a factor.

Dr. William Haddon espoused the theory that unwanted transfers of energy can be harmful (and wasteful) and that a systematic approach to limiting their possibility should be taken. Thus, it is proposed that "a systematic approach" be taken in the design process to limit harmful transfers of energy and exposures to harmful environments.

The questions in Section A relate to Haddon's theory and are presented as general concepts to be considered when using the checklist, for which yes or no answers are to be obtained. They emphasize that two distinct aspects of risk are to be considered in the design process:

- Avoiding, eliminating, or reducing the probability of a hazard-related incident occurring
- Minimizing the severity of adverse results if an incident occurs

A. Introduction: Basic Considerations

1. Can production of hazardous materials or energy be eliminated?
2. Will the amount of the hazardous materials or energy be limited?
3. Can less hazardous materials be substituted?
4. Can hazardous material or energy buildup be prevented?
5. Can release of hazardous materials or energy be slowed down?
6. Can unwanted energy release be separated in space or time from that which is susceptible to harm or damage?
7. Can barriers be interposed to separate the unwanted energy release from that which is susceptible to harm or damage?
8. Will surfaces with which people come into contact be modified to reduce the risk of injury?

B. Designing for Those with Disabilities

1. Do the designs take into consideration the requirements of the Americans With Disabilities Act (ADA)?
2. Are reasonable accommodations made for the disabled?

C. Confined Spaces

1. Have confined spaces been eliminated by design where possible?
2. Are any confined spaces to be permit-required? Refer to 1910.146(c)(1).
3. Have confined spaces been designed for easy of ingress, prompt egress, and where possible, elimination of hazardous atmospheres?
4. Can confined spaces be designed with multiple, large accesses?
5. Are accesses provided with platforms that will support all required personnel and equipment?
6. Will access ports be large enough to permit entry when personnel are using personal protective equipment?
7. Will pipes or ducts limit entry to access ports?
8. Are locations of ladders and scaffolds in the space identified?
9. Are fall protection needs fulfilled (such as anchorage points)?
10. Can the necessary equipment be moved through accesses?
11. Does the design provide for isolation of the confined space from hazardous energy (i.e., electrical, chemical, etc.)?
12. Does the design provide for isolation by valve blocking, spools, double blocks and bleeds, flanges, and flushing connections?

13. Can spaces be designed so that maintenance and inspection can be performed from outside or by self-cleaning systems?

D. Electrical Safety

1. Overall, will the electrical system meet OSHA/NEC standards?
2. Will the system be sufficiently flexible to allow for future expansion?
3. Will emergency power be provided for critical systems?
4. Is grounding adequate?
5. Are ground fault interrupter circuits to be installed where needed? Refer to 1910.304(f)(7).
6. Are grounding connections to piping and conduits eliminated to prevent accumulation of static electricity?
7. Is grounding provided for lightning protection on all structures?
8. Are accommodations made for special purpose or hazardous locations? Refer to 1910.307.
9. Is the design adequate where there may be combustible gases or vapors?
10. Is high-voltage equipment isolated by enclosures such as vaults, security fences, lockable doors, and gates?
11. Are nonisolated conductors such as bus bars on switchboards or high-voltage equipment connections that are located in accessible areas protected to minimize hazards for maintenance and inspection personnel?
12. Where injury to an operator may occur if motors were to restart after power failure, are provisions made to prevent automatic restarting upon restoration of power? Refer to 1910.262(c)(1).
13. Are electrical disconnect switches lockable, readily accessible, and labeled? Refer to 1910.303(f).
14. Are breakers/fuses properly sized? Refer to 1910.303(b).
15. Has the polarity of all circuits been checked? Refer to 1910.403(a)(2).
16. Do electrical cabinets and boxes have appropriate clearances? Refer to 1910.303(g) and (h).
17. Are exposed live electrical parts operating at 50 volts or more guarded against accidental contact by approved cabinets or enclosures, by location, or by limiting access to qualified persons? Refer to 1910.303(g)(2)(i).
18. Are rooms or enclosures containing live parts or conductors operating at over 600 volts, nominal, designed to be kept locked, or provisions made to be under the observation of a qualified person at all times? Refer to 1910.303(h)(2).
19. Are the electrical wiring and equipment located in hazardous (classified) locations intrinsically safe, approved for the hazardous location, or safe for the hazardous location? Refer to 1910.307(b)(1–3).

E. Emergency Safety Systems: Means of Egress

1. Are means of egress adequate in number, remote from each other, properly designated, marked, lighted, and easily recognized?
2. Has emergency lighting be provided for means of egress, and elsewhere where needed?
3. Does the design contemplate emergency lighting where workers may have to remain to shut down equipment?
4. Do means of egress exit directly to the street or open space?
5. Are doors, passageways, or stairways that do not lead to an exit marked by signs reading "not an exit" or by a sign indicating actual use?
6. Does the design provide internal refuge areas for workers who cannot escape?
7. Will reliable emergency power be provided for critical and life-support systems?
8. Will emergency safety showers and eyewash stations be adequate and properly placed?
9. Will adequate first-aid stations, spill carts, and emergency stations be provided?

F. Environmental Considerations (Some Are Operational, Beyond Design)

1. Have waste products been identified and a means of disposal established?
2. Will provisions be made for responding to chemical spills (containment, cleanup, disposal)?
3. Is there an existing spill control plan for chemicals?
4. Have all waste streams been identified?
5. Are adequate pretreatment facilities provided for process waste streams?
6. Will an adequate storage area be available for wastes held prior to treatment or disposal?
7. Will waste storage areas have adequate isolation, or containment for spills?
8. Will hazardous wastes be disposed of at approved treatment, storage, and disposal facilities?
9. Is special equipment or specially trained personnel provided for treatment operations?
10. Has the acquisition of permits been addressed for the treatment or disposal of waste streams?
11. Have state or local requirements for permitting been evaluated and factored into the project?

12. Can the facility meet regulations for reporting spills or the storage of chemicals?
13. Have adequate provisions been made for cleaning the process equipment?
14. Have provisions been made for a catastrophic release of chemicals?
15. Have provisions been made for any necessary demolition and the resulting waste?
16. Have requirements for remediation at the site prior to construction been addressed?
17. Will all feasible measures for waste minimization be implemented?
18. Have the processes that generate air pollution been evaluated for minimization potential?
19. Will adequate air pollution controls be installed (scrubbers, fume hoods, dust collectors)?
20. Have handling and cleaning of air pollution control systems been addressed?
21. Have the processes that generate wastewater been evaluated for minimization potential?
22. Will indoor spills be protected from reaching drains?
23. Will outdoor spills be protected from reaching storm water drains and sewer manholes?
24. Are adequate water disposal systems available?
25. Will pretreatment methods be necessary and provided?
26. Will the discharges of domestic and industrial wastewater be in accord with regulations?

G. Ergonomics: Work Station and Work Methods Design

1. Generally, have material-handling designs considered worker capabilities and limitations, to accommodate the employee population at the 95% level?
2. Do material-handling designs promote the use of mechanical material-handling equipment, such as conveyors, cranes, hoists, scissor jacks, and drum carts?
3. Do design layouts minimize:
   a. constant lifting?
   b. twisting and turning of the back when moving an object?
   c. crouching, crawling, and kneeling?
   d. lifting objects from floor level?
   e. static muscle loading?
   f. finger pinch grips?

   g. work with elbows raised above waist level?
   h. twisting motions of hands, wrists, or elbows?
   i. hyper-extension or hyper-flexion of wrists?
   j. repetitive motion?
   k. awkward postures?
4. Are work stations designed to provide:
   a. adequate support for the back and legs?
   b. adjustable work surfaces that are easily manipulated?
   c. delivery bins and tables to accommodate height and reach limitations?
   d. work platforms that elevate and descend, as needed?
   e. powered assists and suspension devices to reduce the use of force?
5. Has adequate attention been given to:
   a. lighting (to IES requirements)?
   b. heat?
   c. cold?
   d. noise?
   e. vibration?
6. Does the design accommodate the hazards inherent in servicing, maintenance, and inspection?
7. Will there be adequate clearance and ready access to equipment for servicing?
8. Will controls be efficiently located in a logical and sequential order?
9. Will indicators be easy to read, either by themselves, or in combination with others?

H. Fall Avoidance

1. Overall, has the design minimized the need for ladders and stairs?
2. Where work at heights is to be done, has adequate consideration been given to providing work platforms or fixed ladders?
3. Are parapets or guardrails provided at roof edges?
4. Is equipment designed to minimize fall hazards during maintenance, inspection, and cleaning?
5. Does the design provide for fall-arrest measures, such as anchorage points and fall-restraining systems?

I. Fire Protection

1. Overall, in the design, will national and local fire codes, and insurance requirements be met?

2. Will fire pumps, water tanks/ponds, and fire hydrants be adequate?
3. Will risers and post valves be accessible, and protected from damage?
4. Will small hose standpipes be adequate?
5. Will sufficient hose racks be provided?
6. Will special fire-suppression systems be provided?
7. Has containment of fire-suppression water been addressed?
8. Will there be adequate external fire zones?
9. Will emergency vehicle access be adequate?
10. Will flame arresters be installed where needed on equipment vents?
11. Will fire extinguishers be of appropriate types and adequate, and mounted for easy access?
12. Will the design for location of flammables be appropriate?
13. For flammables, will storage rooms and cabinets meet NFC and insurance requirements?
14. For flammable liquid dispensing, will grounding, bonding, and ventilation be adequate?
15. Will fire sensors, pull stations, and alarms be adequate?
16. Are flooding systems designed to provide a predischarge alarm that can be perceived above ambient light or noise levels before the system discharges, giving workers time to exit from the discharge area?
17. Has the project been reviewed by insurance personnel?

J. Hazardous and Toxic Materials

1. In the design process, have all materials in this category been identified?
2. Have the physical properties of the individual chemicals been identified?
3. Have the most conservative exposure limits been established as the design criteria?
4. Has a determination been made to use intrinsically safe equipment?
5. Have Material Safety Data Sheets been obtained for all materials?
6. Are the reactive properties known for chemicals that will be combined or mixed?
7. Have measures been taken to eliminate, substitute for, or minimize the quantities of hazardous chemicals?
8. Does the design emphasize closed-process systems?
9. Will the design properly address all occupational illness potentials, and minimize the need for monitoring, testing, and personal protective equipment?
10. Are storage facilities designed to separate hazardous from nonhazardous substances?

11. Does the design consider the chemical compatibility issues?
12. Have adequate provisions been made for chemical release, fire, explosion, or reaction?
13. Have provisions been made to contain water used in hazardous release control?
14. Are ventilation systems adequate to handle an emergency release?
15. Is the storage of hazardous chemicals below ground avoided?
16. Will storage tank location minimize facility damage or damage to the public in a catastrophic event?
17. Is adequate storage tank diking provided?
18. Will emergency ventilation be provided for accidental releases?
19. For extraordinary releases, will special ventilation, relief, and deluge systems be provided?
20. Will the normal use of chemicals allow operating without personal protective equipment?
21. Will the design of bulk loading/unloading facilities contain anticipated leaks and spills?

K. Lockout/Tagout: Energy Controls

1. In the design process, has adequate attention been given to lockout/tagout requirements to prevent hazardous releases from these energy sources:
   a. electrical?
   b. mechanical?
   c. hydraulic?
   d. pneumatic?
   e. chemical?
   f. thermal?
   g. nonionizing radiation?
   h. ionizing radiation?
2. Are lockout/tagout devices adequate in design and number, readily accessible, and operable?
3. Are lockout/tagout devices standard throughout the facility?

L. Machine Guarding

1. Overall, do the designs prevent workers hands, arms, and other body parts from making contact with dangerous moving parts? Refer to 1910.212(a)(3).

2. Have the requirements of all applicable machine guarding standards of the American National Standards Institute been identified and met?
3. Are safeguards firmly secured and not easily removed? Refer to 1910.212(a)(2).
4. Do safeguards ensure that no object will fall into moving parts? Refer to 1910(a)(1).
5. Do safeguards permit safe, comfortable, and relatively easy operation of the machine? Refer to 1910.212(a)(2).
6. Can the machine be oiled without removing safeguards? Refer to 1910.212(a)(2).
7. Does the design include a system that requires shutting down machinery before safeguards are removed?
8. Are fixed machines soundly anchored?
9. Are in-running nip points properly guarded? Refer to 1910.212(a)(1).
10. Will the design properly address point-of-operation exposure? Refer to 1910.212(a)(3).
11. Are all reciprocating parts properly guarded? Refer to 1910.212(a)(3)(iv).
12. Are all rotating parts properly guarded? Refer to 1910.212(a)(3)(iv).
13. Are all shear points properly guarded? Refer to 1910.212(a)(3)(iv).
14. Are exposed set screws, keyways, collars, etc. properly guarded? Refer to 1910.212(a)(3)(iv).
15. Does the design eliminate the potential for flying chips? Refer to 1910.212(a)(i).
15. Has the potential for any sparking been eliminated? Refer to 1910.212(a)(i).
16. If robots are to be used are they designed to ANSI/RIA R15.06-1999, "The American National Standard for Industrial Robots and Robot Systems—Safety Requirements"?

M. Noise Control

1. In the design process, have maximum noise levels been established that are to be stipulated in specifications for new equipment?
2. Is emphasis given to controlling noise levels through engineering measures?
3. Are the size or shape of rooms and proposed layout of equipment, work stations, and break areas to be evaluated for noise levels?
4. Will workers be separated from noise by the greatest feasible distance?
5. Will barriers be installed between noise sources and workers?
6. Are enclosed control rooms to be provided for operators in areas where the noise is above trigger levels?
7. Are lower noise level processes to be selected, where feasible?

8. Have equipment and work stations been located so that the greatest sources of noise are not facing operators?

N. Pressure Vessels

1. Will all pressure vessels be designed to ASME and insurance company requirements?
2. Will pressure vessels containing flammables or combustibles meet OSHA 1910.106 and NFC standards?
3. Will pressure relief valves be:
   a. correctly sized and set?
   b. suitable for intended use?
   c. directed to discharge safely?

O. Ventilation

1. Have all sources of emission been identified and their hazards characterized?
2. Have ways to reduce personnel interaction with the emission sources (location, work practices) been considered in the design process?
3. Have assessments been made with respect to incompatible emission streams (cyanides and acids, etc.)?
4. Has consideration been given to weather conditions and seasonal variations?
5. Will the design requirements of the ANSI Z9 series, the ACGIH Ventilation Manual, ASHRAE guidelines, and NFPA 45 and 90 be met?
6. Will local ventilation effectively capture contaminates at the point of discharge?
7. Will room static pressures be progressively more negative as the operation becomes "dirtier"?
8. Will the ventilation system provide a margin of safety if a system fails?
9. Will emergency power and lighting be provided on critical units?
10. Will the ventilation equipment be remote and/or "quiet"?
11. Will spray booths and degreasers meet OSHA standards?
12. Will laboratory or contaminated air be totally exhausted?
13. If contaminated air is cleaned and reused, will it meet good safety requirements?
14. Will the make-up air to hoods be clean and adequate?
15. Have flow patterns been established to prevent exposure to personnel?
16. Does the design provide for proper gauging and alarm systems with respect to a sudden pressure drop?
17. Are ventilation controls easily accessible to operators?

P. Walking and Working Surfaces, Floor and Wall Openings, Fixed Stairs and Ladders

1. Will aisles, loading docks, and through doorways have enough clearance to allow safe turns where material handling equipment is used? Refer to 1910 . 22(b)(1).
2. In the aisles, are people and vehicles adequately separated?
3. Are permanent aisles to be marked with lines on the floor? Refer to 1910.22(b)(2).
4. Does the design provide for floors, aisles, and passageways being free from obstruction? Refer to 1910.22(b)(1).
5. Has a logistics study been made to provide safe and efficient flow of people and materials?
6. Will the construction texture of walking surfaces be nonslip?
7. Will the floors be designed to stay dry?
8. Will water and process flows be designed to keep off the walkway?
9. Will the floors be sloped and drained?
10. Will utilities and other obstructions be routed off the walking surfaces?
11. Will the design allow future utility expansion, with added facilities not having to be above and thereby cross floors, and be obstructive?
12. Will designs for floor and wall openings meet the requirements of OSHA 1910.23?
13. Do the designs for fixed stairs, and ladders, meet the requirements of:
    a. OSHA 1910.23, .24, .27?
    b. AFOSH Standard 127-3, -6?

CONCLUSION

There are many references in this chapter to organizations that have, to varying degrees, implemented design and acquisition practices that avoid bringing hazards into the workplace. One of the companies that has implemented extensive procedures to serve that purpose is Intel. Appendix A, duplicated here with permission, is an adaptation of Intel's Equipment Design Philosophy, a philosophy that is supported by a practical management guide on how to avoid bringing hazards into the workplace. It is highly recommended that the reader carefully review this material.

For further information on the plan Intel has put in place for the sake of exemplary design, visit www.intel.com. What Intel requires of suppliers is particularly interesting.

A suitable conclusion for this chapter, which encourages safety professionals to tactfully but aggressively enter the design and redesign processes, is provided by Kathy A. Seabrook in her article entitled, "10 Strategies for Global Safety Management." She counsels safety practitioners to develop a continuous improvement process and look upstream to find ways to positively affect a business through innovation. She cites this example:

> ... A multinational company designed a new product requiring the use of toxic, highly volatile raw material in the manufacturing process. At the time, the design engineer did not understand the implications for use of this raw material; however, the safety and health professional, who is a part of the planning process for all product manufacturing operations, highlighted the additional costs for ventilation, scrubbers and other environmental equipment, blow-out walls and partitions needed to control the risk of using the particularly volatile raw material. Based on this input, the design engineers soon found an alternative raw material that did not need the expensive safety measures.

Seabrook also proposes that safety practitioners develop a process for managing change:

> The concept of change management came from design engineers who had to keep track of changes to engineering plans. Without a process or system to manage these changes, there could be serious ramifications....
>
> This same applies to health and safety issues. Changes in operations ... require an updated risk assessment and associated control measures in the planning stage.

If a safety practitioner looks upstream during the business process, ways can be found to positively impact on results, principally through effective counsel in the planning stages on how to avoid bringing hazards into the workplace.

REFERENCES

Christensen, Wayne C. and Fred A. Manuele, Ed. *Safety Through Design*. Itasca, IL: National Safety Council, 1999.

Design For EHS: Equipment Design Philosophy. Chandler, AZ: Intel Corporation, 1999.

Guidelines for Hazard Evaluation Procedures, 2nd ed. New York: Center for Chemical Process Safety of the American Institute of Chemical Engineers, 1992.

Haddon, William, Jr. "On the Escape of Tigers: An Ecologic Note." *Technical Review*, May 1970.

"Safety of Machinery—Principles for Risk Assessment." European Standard EN 1050, Brussels: Central Secretariat, European Committee for Standardization, 1997.

Seabrook, Kathy A. "10 Strategies for Global Safety Management." *Occupational Hazards*, June 1999.

Appendix A

Design for EHS, Equipment Design Philosophy

MANAGEMENT SUMMARY

Achieving cost effective compliance with environmental, health, and safety requirements has become increasingly difficult with the rapid pace of modern technology cycles and continued globalization within the semiconductor industry. Many companies within the semiconductor industry have worked to create a common set of requirements through the use of SEMI standards such as S2-93 and S8-95.

Unfortunately, significant financial resources continue to be consumed to modify equipment designs because such standards are treated as the basis for EHS compliance audits rather than as design criteria. Companies lose additional financial resources when they must respond to incidents that result from insufficient EHS training or inadequate evaluations during the design and development phases. This document introduces the key aspects of the "Design for EHS" strategy and its benefits.

DFEHS Concept: Integrate EHS requirements and analyses into the tool development cycle, starting at the design concept phase and continuing through each subsequent phase. The system must be closed loop so that lessons learned from previous tool generations are incorporated into new designs and new EHS requirements can be introduced.

Source: Intel Corporation, Corporate Equipment EHS/Capital Equipment Development, June 1999.

Purpose: No aspect of the tool's design, development, installation, operation, maintenance, or service should place any person, facility, or the environment in jeopardy. Equipment manufacturers should assume exclusive ownership for developing and implementing a comprehensive EHS strategy that achieves this purpose. If the DFEHS strategy is properly applied, the equipment manufacturer will address all EHS concerns during the design and development phases, establish safe work practices during equipment development, and automatically document all end user EHS requirements prior to product release.

Competitive Advantages of a "Design For EHS" Model

I. Lowers cost of EHS compliance by almost 60%, improving overall tool development cost
II. Enables an incident- and injury-free manufacturing environment
III. Reduces duration of tool installation and qualification by providing accurate documentation on facility preparation and equipment installation during the customer's project planning
IV. Drives global regulatory compliance
V. Reduces manufacturer's product liability exposure
VI. Reduces manufacturer's time to market for new equipment models

TECHNICAL DISCUSSION

Management's commitment to the design for EHS strategy and participation in ensuring a comprehensive EHS program is critical in achieving a company culture that integrates EHS as a top priority. A culture that emphasizes good EHS practices should produce measurable improvements in the designs of products that employees produce as well as the manner in which they work, both on the company's property and at the customer's site.

Management must define the expectations that project managers, engineers, and other personnel are to use when making decisions on EHS compliance. Each employee must accept the responsibility to ensure the work he or she performs does not lead to a hazardous condition. Several key considerations for developing a design for EHS program are outlined below.

A basic equipment development model showing recommended tasks for each of the primary development phases follows the technical discussion.

Implementation Keys for Successful of a Design for EHS Program

I. Management accepts accountability for the program's success
II. Single point of contact responsible for implementation across all divisions, product lines, etc. This individual is thoroughly linked to management as well as organizations responsible for engineering, quality assurance, reliability, field service, training, etc.

III. Equipment design engineers held responsible for application of the Design For EHS process
IV. Engineering and field service personnel trained to work safely as well as identify and investigate EHS issues at the customer's site
V. DFEHS model applied to full life cycle of the company's products, projects and programs
VI. DFEHS process documents knowledge gained during all projects, product life cycles, etc.

Fundamental Components of a Design for EHS Program

I. Documented EHS programs
II. Documented equipment development procedure integrating EHS milestones
  A. Defined roles and responsibilities
  B. Reference library (e.g., standards, codes, checklists, and other design "tools" for engineers)
  C. Established milestones within the development cycle (e.g., EHS Roadmap, design reviews)
  D. Consistent hazard analysis techniques (e.g., failure modes and effects analysis)
  E. Consistent risk assessment techniques (e.g., SEW S10–1296)
  F. Established hazard control hierarchies (e.g., personnel safety, pollution prevention)
  G. Company policy defining EHS documentation expected from project teams
  H. Database where lessons learned from other equipment models are documented
  I. Project updates to a management review board (monitors status, reviews key decisions, etc.)
  J. Process checks
III. Internal program audits

BASIC DESIGN FOR ENVIRONMENTAL, HEALTH AND SAFETY MODEL TASKS FOR EACH PHASE OF EQUIPMENT LIFE CYCLE

Concept Phase

- Evaluate new product idea
- Compare functionality needs against available technologies
- Determine critical components and subsystems required

- Identify environmental, health, and safety risks associated with the proposed technology
- Prepare list of design options with the suitable EHS characteristics for evaluation during feasibility study

Concept Feasibility

- Review design options against EHS database to ensure lessons learned in previous generations are given proper consideration
- Evaluate the quantity of chemicals and natural resources expected to be consumed by the process technology
- Establish EHS requirements necessary to meet goals for EHS performance and code compliance
- Conduct an initial hazard survey to identify potential EHS issues during installation, operation, maintenance, service, and decommissioning
- Make design recommendations to project team
- Develop EHS roadmap which integrates EHS requirements as key milestones in the equipment development schedule

Development of Enabling Hardware

- Communicate to the design engineers how the EHS roadmap fits into the equipment development schedule
- Provide the design team with training on EHS requirements
- Integrate EHS requirements into project success criteria
- Comprehend impact of variations in customer configurations, facility designs, and jurisdictional requirements
- Conduct preliminary hazard review and code analysis: ergonomics, safety, electrical, clearance zones, materials of construction, etc.
- Evaluate methods to reduce the consumption of chemicals and other natural resources
- Include EHS in project design reviews. Add any lessons learned to EHS database
- Inform suppliers of components and subsystems of the EHS requirements and integrate them into purchasing specifications
- Develop procedures to safely control all hazards associated with operation, maintenance and service and prepare first draft of manuals. Add any lessons learned to EHS database
- Develop EHS test plan

Development of Pilot Tool

- Evaluate and test all interfaces, subsystems, components, schematics, interlocks, and emergency systems according to EHS test plan

- Conduct detailed hazard analysis of facilitation, installation, operation, calibration/testing, and maintenance activities
- Develop documentation for facility preparation and equipment installation
- Update EHS information in the manuals for operation, maintenance and service based on evaluation of results
- Collect environmental emissions and utilities data and assess against goals
- Define training requirements for all job functions
- Perform EHS and code compliance evaluation
- Implement design modifications and revise documentation as necessary to achieve EHS and code compliance goals
- Track all issues to closure

Manufacturing Readiness

- Develop training courses for internal and end user personnel
- Validate closure of EHS roadmap and all identified compliance issues
- All components and subsystems are listed or recognized by Nationally Recognized Testing Laboratory (NRTL) and/or validated by third party review
- Define standard equipment configuration (may impact tool installation and jurisdictional requirements, or vice versa)
- Evaluate all configuration options against EHS requirements to ensure full compliance. Correct design(s) as necessary
- Comprehensive EHS evaluation by qualified professionals (internal or third party), including operations and service manuals
- Access any discrepancies identified and take corrective action
- Obtain all compliance documentation and supporting data: Technical Construction File (TCF), SEMI S2/S8 reports, compliance letters, tool manuals, required certifications, etc.

Product Release/Full Manufacturing

- Conduct a final comprehensive EHS systems check and update environmental emissions data
- Make final revisions to installation documentation and associated safety procedures
- Notify customer of install, layout and facilitation requirements needed to maintain EHS compliance and meet customer's special requirements
- Publish final revisions of operating and maintenance manuals
- Offer training courses to internal and end user personnel
- Conduct post mortem of project and add lessons learned to EHS database

Installation, Operation and Maintenance

- Establish closed loop feedback system with field service personnel and customers to ensure quick and proper disposition of newly discovered EHS issues
- Engineering and customer support must give priority to investigating all EHS issues and implementing corrective action
- Establish a proactive product safety notification system for all customers
- Change control procedures must be established to review and implement corrective actions in a controlled manner when EHS issues are discovered
- Lessons learned from EHS issues discovered during the tool's operational life should be added to the EHS database for use in design projects

Equipment Decommissioning

- Identify and assess safety and health hazards resulting from equipment demolition
- Identify hazardous or regulated wastes, contaminated components, etc. which present potential environmental issues
- Determine the environmental regulations which apply to the situation and specify best known treatment/disposal methods
- Provide customers with procedures to safely perform demolition, decontamination, and disposal of equipment and any associated waste
- Capture lessons learned during decommissioning in the EHS database for use in designing the next generation tool

8

Task Analysis: For Productivity, Cost Efficiency, Safety, and Quality

INTRODUCTION

Without question, the current business climate requires a higher level of management effectiveness, a part of which is a stringent measurement of the contributions of individuals toward achieving entity goals. Demands to reduce costs and improve productivity and quality are never-ending.

Safety practitioners with whom the author has had recent discussions perceive a need to bring additional value to their companies, and to be recognized as problem solvers whose counsel is desired because they are successful participants in achieving management goals. One mark of success is being sought out for advice, which results from activities that convince internal customers the safety practitioner is a resource that will help them achieve their goals. It seems that "being on the manager's page" is a commonly used phrase in the language of safety practitioners.

Job safety analysis methodologies have been available for decades. They have been used effectively in some companies, and tried and abandoned in others because of perceived excessive time requirements.

Nevertheless, when we reflect on the current business climate and consider the probability of achieving more effective safety management results, anecdotal data indicate that safety practitioners should consider the benefits of transforming long-established job safety analysis methods into a task analysis system that incorporates practical combinations of ergonomics, productivity, cost efficiency, and quality.

RATIONALE

Several factors give impetus to such a proposal. They include a review of data on the applications of ergonomics, extensive studies of injury and illness investigation reports, and successes in combining quality considerations within what originally were job safety analysis systems.

The author has been significantly influenced by the immense body of literature indicating that applications of ergonomic principles for occupational risk reduction often also result in improved productivity and lower operating costs, and sometimes improved quality.

Such anecdotal data are voluminous, and much is to be gained by revising our task analysis systems to reflect on what we have learned. For quite some time, the author had been using a simple definition of occupational ergonomics that corresponded to what was common in the literature: Occupational ergonomics is the art and science of designing the work to fit the worker. If we reflect on the reality of the data collected, the following definition may now be used:

> Occupational ergonomics is the art and science of designing the work to fit the worker: to achieve optimum productivity and cost efficiency, and minimum risk of injury.

Note the order in which productivity, cost efficiency, and risk of injury are placed. Emphasizing the possible productivity and cost-efficiency gains from applied ergonomics is a different approach from that found in most of the prominent ergonomics literature. And, it is to the advantage of safety practitioners who seek accomplishment and recognition to do so.

Consider the following example. A safety professional had difficulty convincing a plant manager that ergonomics studies should be conducted at an operation because of its injury and illness experience. Reluctantly, the plant manager conceded and assembled a study team. When the recommended ergonomics improvements had been made, production personnel made cost-benefit analyses and concluded that not only had risks been reduced, but productivity and cost efficiency had also been favorably impacted. From that day forward, the safety professional stressed the possibility of improving productivity and cost efficiency, and sometimes quality, in his proposals for ergonomics studies. And, his counsel was in greater demand.

We can conclude from the lessons learned in the last several years that it should be the exception when an ergonomics task analysis is proposed only for safety purposes: We should also be addressing productivity and cost efficiency.

Other factors have influenced such a point of view. As we became more astute in incident analysis for causal factor determination, it was often concluded that the operations and work methods being analyzed were not always the most productive or efficient. And, as employee participation expanded in recent years through employee involvement and empowerment initiatives, it frequently occurred that employees recommended less risky, easier, and more productive ways than prescribed methods to do the work.

The task analysis method proposed here requires employee involvement, because of the benefits that can and do derive from their participation.

TAKING THE APPROACH THAT WORKS

What is proposed is that safety practitioners consider a revision of job safety analysis/job hazard analysis methods to become a task analysis system that addresses safety, ergonomics, productivity, cost efficiency, or quality—choosing whatever combinations have the best possibilities for acceptance and success, after considering operational needs and opportunities. In so doing, an assessment is required of the potential developments and gains that can be achieved.

A few safety professionals have developed such systems. As an example, in a situation where a safety professional broadened his job safety analysis procedures to include quality considerations, that was the best approach to take because of the needs in his organization, and it worked. His system addresses safety risks and quality specifications in an activity flow chart, and considers process alternatives and preventive actions. In the examples reviewed, it was obvious that consideration was given to the best and most efficient way to accomplish a task.

Giving the extended system a new name, relating it to management goals, may also help in its marketing: e.g., TAPES for task analysis for Productivity, Ergonomics, and Safety; or TAPCERA, task analysis for Productivity, cost efficiency, and risk assessment. Whatever might lead to its widespread acceptance is worth considering.

JOB OR TASK?

Thus far in this chapter, the words "job" and "task" have been frequently used. The term job has several meanings; e.g., it may designate an occupation title, or a specific task assignment. The term task is more limiting in scope; often, it is a part of one's job and the factor that is to be analyzed. In *Practical Loss Control Leadership*, 2nd edition, Bird and Germain give this definition for a task:

> A task is a segment of work which requires a set of specific and distinct actions for its completion (150).

This definition is proposed for use in an extended task analysis system.

REVISING OUR MARKETING APPROACH

Unfortunately, in the highly competitive business climate that has evolved, having operating personnel perform job safety analyses may not be given a high priority. We may not like having job safety analysis given a low priority, but that is often the reality of an organization. Also, we must consider that doing job analyses for safety purposes alone can be perceived as separating safety from the core business.

The following comments, taken from *Practical Loss Control Leadership*, 2nd ed., reflect the real world:

> This approach (job safety analysis) usually examines the work only from the perspective of safety and health. It has resulted in safer work. But it has also resulted in duplication of effort and paperwork, and confusion with safety procedures, quality procedures, efficiency procedures, etc. Because procedures which deal only with safety are not always perceived to be related to the primary purpose for doing the work, they tend to get ignored in the face of other pressures (149).

Our marketing opportunities can be greatly enhanced if we produce a task analysis system that is perceived as having a more direct impact on operational needs and is consistent with management goals. Management talks the language of productivity, cost efficiency, and quality, and extending our task analysis systems as proposed here gives safety practitioners greater opportunity to "get on the manager's page."

Safety practitioners often become frustrated by management decisions that are perceived as indicating a lack of support. On a given day, managers have before them multiple proposals made by the staff that require spending money. And, the reality is that the proposals made by safety personnel are in competition with all other proposals. The suggestions here may help move the recommendations made by safety practitioners higher up on the ladder of priorities.

Under no circumstance is it suggested that the thoroughness of the safety and ergonomics aspects of a combined task analysis be diminished in purpose or scope, or be less than professional. It is proposed that completing such task analyses more often better serves safety management purposes.

A REVIEW OF THE LITERATURE

A review was made of 35 safety-related texts to determine what other authors have said about relating job safety analyses to productivity, cost efficiency, or quality. Such references are few and mostly very brief. Roger L. Brauer wrote this in *Safety and Health for Engineers*:

> A job safety analysis can be completed concurrently with other forms of task analysis common to industrial practice (463).

Edward L. Grund in *Lockout/Tagout: The Process Of Controlling Hazardous Energy* made this comment about task analysis:

> This technique can break down lockout/tagout procedures into a series of measurable steps in which each task is analyzed for its effectiveness, safety, and necessity (261).

Willie Hammer referred to effectiveness and efficiency when writing about procedure analysis in *Occupational Safety Management and Engineering*, 3rd ed.:

> A procedure analysis is a set of instructions for sequenced actions to accomplish a task.... Analyses must ensure that the procedures are not only effective and efficient but safe (134).

The National Safety Council has a training program entitled, "Job Safety Analysis: Identifying and Controlling Hazards and Managing the Process." In Module 1 of the Instructor's Guide, the following is stated:

> The process of developing a job safety analysis enables employees to identify task-related hazards and to establish safer work procedures that management endorses, improving work place safety, productivity, and quality (1).

William M. Montante, in "An Ergonomic Approach to Task Analysis," wrote this about analyses benefiting the overall business system:

> An ergonomic approach to task analysis (ETA) expands the scope of traditional methods by evaluating, step-by-step, how the human operator interacts with machine, product, system and work environment. This analysis helps determine whether task demands match—or exceed—human capabilities. Regardless of analysis approach, results can substantially benefit the overall business system (18).

Other authors have written extensively on the desirability of having job analyses encompass all aspects of the work to be done. In Chapter 5 entitled, "Proper Job Analysis and Procedures," in *Management Guide to Loss Control*, published in 1974, Frank Bird pleaded extensively to have job analyses address more than safety. In *Practical Loss Control Leadership*, 2nd edition by Frank E. Bird, Jr. and George L. Germain, published in 1985, more extensive treatment of Bird's original work was given. Note: the term "job analysis" became "task analysis."

The following are the premises on which Bird's chapter entitled, "Proper Job Analysis and Procedure," is based. They were written 27 years ago. Although they were sound then and remain so now, it is unfortunate that few safety practitioners have adopted them:

> The proper job analysis and standard job procedure chapter is based on the concept that all elements of a worker's job, such as quality, production, safety and health, are inseparable. Any one or all can affect the others, and to consider them as separate elements when teaching a worker to do his job is to invite the confusion and misunderstanding that leads to downgrading incidents.
>
> It just isn't possible to do a job efficiently if all aspects of the job are not considered as part of the standard procedure to get it done. Efficiency requires the best use of people, equipment, materials, and environment, all working together to produce optimum results (60).

Bird proposed that there be one standard task procedure which encompasses all aspects of the task and that such a procedure could best be written if all the factors of the task were included in the task analysis process. That is also the premise of this book. The method proposed here begins with the job safety analysis or job hazard analysis procedures with which most safety practitioners are familiar.

GETTING STARTED

Those who choose to extend job safety analysis to include aspects of ergonomics, quality, or productivity and cost efficiency would do well to spend some time with the chapters previously referenced in *Management Guide to Loss Control* and *Practical Loss Control Leadership*. They are thought-provoking.

In the National Safety Council's *Accident Prevention Manual for Business and Industry: Administration & Programs*, 11th edition, a traditional dissertation on job safety analysis is given (134). But, that traditional approach has been extended to include considerations of productivity and quality in its training program entitled, "Job Safety Analysis: Identifying and Controlling Hazards and Managing the Process." Also, safety practitioners may want to consider enhancing their knowledge by taking one of several courses available on job safety analysis.

THE TASK ANALYSIS PROCESS

It is not the intent of this paper to review task analysis procedures in detail. References previously cited are good resources for such a review. The section on task analysis in the *Human Factors Design Handbook*, 2nd edition by Woodson, Tillman, and Tillman (760) is also recommended. Comments will be made here on some aspects of the Bird and Germain system and that advocated by the National Safety Council.

Bird and Germain present "... a systematic, practical approach to preparing and using Task Procedures and/or Practices" in seven stages:

1. Inventory occupations
2. Inventory all tasks within each occupation
3. Identify the critical tasks
4. Analyze the critical tasks
   (a) Break tasks down into steps or activities
   (b) Pinpoint loss exposures
   (c) Make an improvement check
   (d) Develop controls
5. Write procedures of practices
6. Put to work
7. Update and maintain records (150)

In the National Safety Council training program (Module 2), the following are the steps in the job safety analysis process:

- Occupational job inventory (job task inventory)
- Job (task) selection
- Sequence of basic job steps: using a standard form, break down the job task into observable steps
- Hazard identification
- Hazard control measures
- Job safety analysis review and approval
- Implementation (6)

There are similarities and differences in these two outlines. Each proposes developing occupational and task inventories.

CONCERNING THE DEVELOPMENT OF JOB INVENTORIES

As the discussion of developing inventories progresses here, the focus is on safety-related priorities because that is the field of endeavor for safety practitioners. However, it should be understood that our purpose is to extend the analysis system into other areas and that follow-up discussion with production and quality-control personnel may result in the selection of tasks by priority that are also of interest to them. As the task analysis is extended into other areas, revisions would be made in analysis systems, which might include adding appropriate columns in analysis forms.

Bird and Germain state the following:

> The first step in developing a comprehensive task inventory is to produce a complete list of all occupations, and the second step is listing all the tasks performed within each occupation (151–152).

The National Safety Council proposes (Module) this:

> The first step is to develop a comprehensive job task inventory for every occupation or job title in the organization (6).

Developing an inventory of *all occupations* may be manageable, but developing an inventory of *all tasks* is daunting. No organization running with the slim staffs that are now typical could manage that, even if it was desirable. Consider the probable reality in an auto parts manufacturing plant: about 1,000 employees, each performing a minimum of 10 tasks. Developing an inventory of every task in such an organization could be so time-consuming that it would never be completed.

In both the Bird and Germain and National Safety Council systems, it is proposed that the critical tasks be selected for analysis, and it seems that doing this first is more practical and attainable.

Identifying Tasks to Be Analyzed

Without question, some tasks are inherently more hazardous than others. If we consider what is best both for employees and employers, those tasks should be given priority for analysis if they present the potential for an incident occurring that may result in serious injury, illness, property damage, or environmental damage.

In determining what tasks are to be analyzed, first, the following would be considered:

1. Actual incident history: OSHA recordable and lost work-day cases, and workers compensation history
2. Incidents that have occurred which might have resulted in serious injury or damage, under slightly different circumstances
3. Input from workers on the tasks they perceive to be hazardous
4. Hazardous, nonroutine tasks (e.g., maintenance and troubleshooting)
5. Other known highly hazardous tasks
6. New tasks with which workers may not be familiar

Throughout the process, judgment and experience will prevail. Seeking input from those who do the tasks, the workers, and encouraging their involvement is a necessity. Actual incident history, obviously, provides an indication of which tasks should be given priority. However, importance should also be given to incidents that could have resulted in serious harm or damage, the so-called "near-miss" event, and tasks about which workers express concern. To identify those, it is vital that a modification of the Critical Incident Technique be applied. This involves worker input. William E. Tarrants, in *The Measurement of Safety Performance*, wrote the following:

> In applying the technique, an interviewer questions a number of persons who have performed particular jobs within certain environments and asks them to recall and describe unsafe errors that they have made or observed, or unsafe conditions that have to come to their attention (304).

The intent is to have workers identify as many critical incidents as they can recall, especially those that did not result in injury or damage. In the process, the interviewer should seek to identify tasks from which low-probability incidents might occur that could result in serious injury or damage. It is in both the worker's and the employer's best interest that a good job be done of trying to anticipate low-probability and high-consequence events before they occur.

Another resource that can be helpful in identifying tasks to be analyzed is *Profitable Risk Control* by William W. Allison (See especially Chapter 5, entitled, "How to Identify Costly Risks Before Costly Losses Occur," 44). Unfortunately, serious injuries too often occur while workers are doing non-routine tasks, and they deserve special attention.

For safety purposes, what is sought here is the identification of hazards in a broad sense. Hazards encompass all aspects of technology and activity or inactivity that produce risk. A hazard is defined as the potential for harm: Hazards include the characteristics of things, and the actions or inactions of people.

Although identifying tasks to be analyzed starts with hazardous operations, it should be kept in mind that the end product developed is to also address a combination of productivity and cost efficiency, quality, or ergonomics.

Setting Priorities

After developing a list of tasks to be analyzed, the next step would be to prioritize them. This is not necessarily a scientific process: The concerns of employees, real or unreal, sometimes will determine what tasks are analyzed first. In the priority-setting process, an adaptation of a risk assessment method, such as that contained in the National Safety Council's Job Safety Analysis Training Program (Module 2) would be helpful, but such a risk assessment system should be used with caution.

Incidents that could have catastrophic results (fatalities, loss of facility, environmental damage) but with low occurrence probability, with only a few people performing the related tasks, are subordinate in the decision-making process if the risk scores are the only priority determinants. Sometimes, good judgment will dictate that incidents having the potential for severe injury or damage be placed at the top of the priority list, even though the scoring for them is low.

The following excerpt from the National Safety Council's Job Safety Analysis Training Program is included here with permission. These are the column headings in its risk assessment form:

Sample Ranking System

Job Task	Severity	Exposure	Probability	Total

In the first column, each identified task would be entered. For each task, numerical values (point values) are then given:

- Potential for severity of injury, property damage, or production interruption
- Frequency of exposure
- Incident probability

In the last column, the numerical entries are totaled. The Council cautions that the numerical scorings it suggests are "arbitrary." This implies that the developer of a rating system should use good judgment in selecting rating category descriptions and the numerical scorings to be ascribed to them. For all categories, input from knowledgeable people, including workers who perform the tasks, must be gathered when selecting numerical scorings. These are the example scorings as given in the National Safety Council program.

Severity of Harm or Damage. Assumptions would be made as to the worst credible outcome of an incident in selecting a numerical rating for severity:

0	Negligible	Probably no injury or illness; no production loss; no lost work days
1	Marginal	Minor injury or illness; minor property damage
2	Critical	Severe injury or occupational illness with lost time; major property damage; no permanent disability or fatality
3	Catastrophic	Permanent disability; loss of life; loss of facility or major process

Exposure. Exposure considers the number of employees who perform a task, the number of times an individual employee performs the task, or both. A sample ranking classification follows:

0	A few employees perform the task up to a few times a day.
1	A few employees perform the task frequently.
2	Many employees perform the task frequently.

Probability. Determining the likelihood of a hazards-related incident occurring is a judgment call. Rarely will there be a satisfactory statistical base that can be used as a reference. Probability is expressed in probable occurrences per unit of time or operations or activity (e.g., per shift, number of times the task is performed, per year.) Determinations are influenced by a number of factors, such as the hazards associated with the task, difficulty in performing the job, the complexity of the job, and whether the work methods are error-provocative. The suggested probability scorings are as follows:

0	Low probability (extremely unlikely, to unlikely but possible to occur)
1	Moderate probability (likely to occur sometime)
2	High probability (will occur several times, possibly frequently)

As the National Safety Council program indicates, after numerical point values are given to each task for severity, exposure, and probability, the point values are added to produce a risk score. A total score can be as low as 0 or as high as 7. So, each task is given a risk ranking from which judgments can be made in setting priorities for task analysis.

This system does not produce absolute guides: Risk assessment is as much art as science. To satisfy employee concerns, it may be prudent in actual practice to analyze certain tasks with lower scores before tasks with higher scores are analyzed. And, as was previously stated, it may be necessary to pay special attention to low probability incidents that have serious injury or damage potentials.

In addition, several tasks may be given the same rating and further judgmental choices will be necessary in the priority-setting process.

Encouragement is given to applying a risk assessment method in prioritizing tasks to be analyzed, *and* to the development of the criteria descriptions and scoring systems that are suitable to particular operations.

EXTENDING THE TASK ANALYSIS PROCESS

Now that tasks have been identified for analysis and priorities have been set, task analyses follow. To extend the task analysis process to include ergonomics or productivity and cost efficiency, or quality, the safety practitioner needs to understand the conceptual significance of what is to be undertaken. The conventional job safety analysis system would be rewritten, the safety practitioner would convincingly make a few analyses using the extended system, and a training program to indoctrinate others in the revised system would be undertaken.

This has to be done on a "designer" basis to assure that the system practically addresses needs and opportunities and can successfully overcome barriers in individual operations. In designing an extended task analysis system, consideration should be given to the following extrapolations, taken principally from the many resources the author reviewed. When determining which of them are to be included in an extended task analysis system, aspects of productivity, cost efficiency, quality, and safety should be considered.

- What is this task to accomplish?
- Is this task necessary: Can it be eliminated?
- Is there a more efficient way to perform this task?
- Is the equipment appropriate to the task?
- Would efficiency be improved if other equipment was provided?
- Is the equipment being properly maintained?
- Is the equipment being properly used?
- Do environmental conditions negatively affect task performance (illumination, noise levels, air contaminants, temperature extremes, ventilation, etc.)?

- Would revisions in workplace layout and configuration improve task performance effectiveness?
- Is there one, written, easily understood procedure that is appropriate for all aspects of the task?
- Is staffing for the task adequate (number of people, skills)?
- Is the task structured to encourage or require deviation from established procedures?
- Do the methods prescribed to do the task encourage risky behavior?
- Could employee errors be attributed to workplace or work methods design?
- Is the design of the workplace or the work methods overly stressful?
- Does the task require excessive physical exertion, or awkward body postures, extreme reaching, bending, twisting, etc.?
- Is the design of the workplace or the work methods error-provocative?
- Can the work methods be redesigned to make errors less likely?
- Do task requirements induce fatigue?
- Are task requirements difficult or irritating?
- Is there easy access for maintenance tasks?
- Are the storage and use of smaller quantities of hazardous substances practicable?
- Can hazardous substances be replaced with less hazardous substances?

COMPUTERS AND DATA BANKS

Obviously, all sorts of possibilities arise for the writing of computer software programs for an extended task analysis system. Beyond that, though, is the importance of sharing the lessons learned within an organization where similar tasks are performed. This suggests the importance of developing data banks and access to them. Of special significance is the ability to then share analyses with facilities throughout the world so that each one does not have to develop analyses for similar tasks.

CONCLUSION

The author believes that the ideas presented here offer safety practitioners a great opportunity to improve the effectiveness of their safety management endeavors while, at the same time, positively affecting the results by which management is frequently measured. The principle measure of success of an extended task analysis system will be the extent to which operations people seek its use.

REFERENCES

Accident Prevention Manual for Business and Industry: Administration & Programs, 11th ed. Itasca, IL: National Safety Council, 1997.

Allison, William W. *Profitable Risk Control*. Des Plaines, IL: American Society of Safety Engineers, 1986.

Bird, Frank Jr. *Management Guide to Loss Control*. Loganville, GA: Institute Press, 1974.

Bird, Frank E. Jr. and George L. Germain. *Practical Loss Control Leadership*, 2nd ed. Loganville, GA: Det Norske Veritas (U.S.A.), 1996.

Brauer, Roger L. *Safety and Health for Engineers*. New York: Van Nostrand Reinhold, 1990.

Grund, Edward L. *Lockout/Tagout*. Itasca, IL: National Safety Council, 1995.

Hammer, Willie. *Occupational Safety Management and Engineering*, 3rd ed. Englewood Cliffs, NJ: Prentice-Hall, 1985.

Job Safety Analysis: Identifying and Controlling Hazards and Managing the Process—A Training Program, Modules 1 and 2. Itasca, Il: National Safety Council, 1994.

Montante, William M. "An Ergonomic Approach to Task Analysis." *Professional Safety*, Feb. 1994.

Tarrants, William E. *The Measurement of Safety Performance*. New York: Garland Publishing, 1980.

Woodson, Wesley E., Barry Tillman, and Peggy Tillman. *Human Factors Design Handbook*, 2nd ed. New York: McGraw-Hill, 1992.

9

Risk Scoring Systems

INTRODUCTION

All of the risk assessment matrices discussed in Chapter 6, "A Primer on Hazard Analysis and Risk Assessment," are two-dimensional. Risk levels are established based on determinations of the *probability* of event occurrence and the *severity* of harm or damage that could result.

In two-dimensional risk assessment matrices, the risk levels are deduced from the combinations of the gradations of event probability and severity of outcome. Management risk decision options for remedial action or for risk acceptance can be established for each of the risk categories shown in the matrices.

For the purpose of illustration, and to serve as a reference when considering the risk scoring systems to be discussed here, a rather ordinary two-dimensional risk assessment matrix is shown in Table 1. A display of management decision levels, relating to Table 1, is given in Table 2.

TABLE 1 Risk-Assessment Matrix

	Severity of Consequences			
Occurrence Probability	Catastrophic	Critical	Medium	Minimal
Frequent	High	High	Serious	Moderate
Likely	High	High	Serious	Moderate
Occasional	Serious	Serious	Moderate	Low
Remote	Serious	Moderate	Moderate	Low
Improbable	Moderate	Moderate	Low	Low

TABLE 2 Management Decision Levels

Risk Category	Remedial Action or Acceptance
High	Operation not permissible.
Serious	Remedial action to have high priority.
Moderate	Remedial action to be taken in appropriate time.
Low	Risk is acceptable: remedial action discretionary.

Making two-dimensional risk assessments has been a broadly applied, but not exclusive, practice for many years. In the process, translating gradations within the event probability and severity of outcome categories into precise numbers, and thereby calculating risk scores from them, is not necessary. However, the world of risk assessment is in transition. Some systems now in use are three- or four-dimensional and require numerical risk scorings. Safety practitioners can expect that variations of risk scoring systems will proliferate and become the norm, over time.

Two-dimensional risk assessment systems are not obsolete. They will continue to serve well in many situations. Often, a two-dimensional system will be selected because it works within a company and its culture. The intent here is to address recent developments and explore what safety practitioners ought to know about them. For those purposes, this chapter reviews:

- Origins of the transitions taking place
- Cautions to be considered in using three- and four-dimensional systems
- Three-dimensional, numerical risk scorings systems in use
- Extended three-dimensional risk scoring systems that include a method to justify the risk amelioration costs in relation to the amount of risk reduction to be attained
- Four-dimensional risk-scoring system, now having limited use, but for which broader use has been proposed

To conclude this chapter, a model risk scoring system will be presented that can be applied for all risks of injury or damage which a safety professional may encounter.

ORIGINS OF THE TRANSITIONS IN RISK ASSESSMENT

In the education of engineers, a passion for quantification and the application of numbers evolves. Engineers become comfortable with statistical measurements and expect things to be measured. Engineering texts still quote Lord Kelvin who wrote the following over 100 years ago:

> When you can measure what you are speaking about, and express it in numbers, you know something about it: but when you cannot measure it, when you cannot express it in numbers, your knowledge is of a meager and unsatisfactory kind.

Some disagree with Lord Kelvin's absolutist statement, such as Vernon L. Grose. In his book *Managing Risk: Systematic Loss Prevention for Executives*, Grose expressed concern over the use of "numerology" (263) in making risk assessments and cautions us about the:

> ... increasingly common abuse of mathematical statistics by "numericalizing without data" to obtain risk probabilities, and the deception-by-numbers that lulls executives into a false sense of security (259).

Nevertheless, the influence of engineers in risk assessment practices will be felt. Their passion for numerical precision and measurement encourages the use of quantified methods.

As an example, the use of quantitative Failure Modes and Effects Analyses (FMEAs) has brought about changes in subjective risk assessment practices. In some industries, quantitative FMEAs are now required to meet quality assurance requirements. This is a fairly recent development. Engineers mostly complete these FMEAs. In at least two industries, the use of three-dimensional numerical risk scoring systems for FMEAs has emerged.

Similarly, some engineers meeting the recently established risk assessment requirements placed on manufacturers who sell machinery to Europe are using numerical risk scoring systems. One such system in use for that purpose is four-dimensional. Also, a three-dimensional risk scoring system is in use in a segment of the heavy machinery manufacturing industry. It was introduced by engineering personnel to meet the demands for product safety.

These developments are indications of the transitions taking place in risk assessment, from two-dimensional methods to newer techniques.

SCOPING THE VARIATIONS IN RISK SCORING SYSTEMS

There are substantive variations in the elements to be scored in the three-dimensional and the one four-dimensional risk scoring systems reviewed here. To commence the discussion, we will reexamine the elements in the risk scoring system of the National Safety Council's Job Safety Analysis Training Program. That system was discussed in Chapter 8, "Task Analysis for Productivity, Cost Efficiency, Quality, and Safety." Numerical values to reach a risk score are assigned to:

- Potential for severity of injury, property damage, or production interruption
- Frequency of exposure
- Incident probability

Other three-dimensional systems retain the probability and severity categories, but substitute other elements to be scored, such as criticality, detectability, and vulnerability.

Two three-dimensional systems that also use probability, severity, and frequency of exposure as elements to be ranked also propose an efficacy test in which the cost of the risk amelioration proposed is related to the amount of risk reduction to be attained.

One four-dimensional system is in use, the origin of which seems to be European. Variations of it have appeared several times. Its elements are likelihood of occurrence, frequency of exposure, degree of possible harm, and number of persons exposed to the hazard.

For each of these systems, the words used to designate the categories to be rated may be the same, but their definitions may differ. Additional comments about the specific risk-scoring systems in use will be made later in this chapter.

NEED FOR CAUTION, AND PERCEPTIVE EVALUATION

Although three- and four-dimensional risk scoring systems will likely become prevalent, there are several reasons to be cautious in their use. Users of such systems should be inquisitive and cautious concerning their content, the meanings of the terms adopted, the numerical values applied to the gradations within elements to be scored, and how they are applied in determining risk levels:

- Risk scorings begin with subjective judgments on the individual elements to be considered, and those subjective judgments are translated into numbers, not followed by any qualifying statements. What starts out as judgmental observations become finite numbers, which then leads to an image of preciseness.

  Further, those numbers are multiplied or totaled to produce a risk score, giving the risk assessment process the appearance of having attained the status of science. In reality, the risk assessment process is as much art as science.
- The numbers assigned to the elements to be scored are entirely judgmental, have no basis in fact or good science, and vary for the same subject in different systems. There are no universally applied rules in assigning values to elements being scored. All values in all risk scoring systems reflect the experience and views of those who create the system.

  The following example illustrates the point. Mention was made earlier in this chapter of the risk scoring system in the National Safety Council's Job Safety Analysis Training Program. In that system, scores for each of the three elements began with a 0. The maximum score that could be attained was 7. In the National Safety Council's *Accident Prevention Manual for Business and Industry: Administration and Programs*, 11th ed., an almost identical risk scoring system appears. It uses the same scoring elements, and the descriptors for the gradations within each element are identical. But, the scorings begin at 1, and the maximum score can be as high as 10 (77).

- Frequency of exposure has historically been one of the elements considered in determining event probability. However, giving frequency of exposure its own multiplier, separate from and in addition to a probability multiplier, will diminish the needed emphasis on the severity of harm or damage that could result from an event.
- The meanings of the terms probability, likelihood, frequency of exposure or endangerment, and severity commonly used in risk scoring systems are not consistent. Also, it may be that within a system, a term may appear more than once and have different usages; and consistency in the meanings of terms used one way initially in the system may not be maintained.
- If risk scoring systems are applied rigidly whereby a higher score indicates a greater risk than a lower score, and the scores are not considered in light of the employee, community, social, and financial concerns of the organization, its best interests may not be well served. This applies especially to low-probability events that may have severe consequences, for which risk scores may be low.

THREE-DIMENSIONAL RISK SCORING SYSTEMS

Since the risk scoring system in a module for the National Safety Council's Job Safety Analysis Training Program is covered in depth in Chapter 8, it will not be reviewed again here. It is not an overly complex system and can be used with good effect if the cautions described are observed.

In the discussion of Failure Mode and Effects Analysis (FMEA) in Chapter 8, it was said that extensions have been made to the basic FMEA method to make it three-dimensional. For example, the publications on FMEA by a consortium of automobile manufacturers and by a group of semiconductor manufacturing companies add a detection criteria to the FMEA process. Discussions of these publications follow. Then, comments will be made on a hazard analysis system used within a segment of the heavy equipment manufacturing industry. Finally, reviews will be given of the two three-dimensional risk scoring systems mentioned earlier that include a justification mechanism relating the cost of risk amelioration to the amount of risk reduced.

Automobile Industry FMEAs

Daimler-Chrysler, Ford, and General Motors are the companies represented in the Automotive Industry Action Group (AIAG). FMEA teams in those companies, working under the auspices of the American Society for Quality and AIAG, developed a reference manual entitled, *Potential Failure Mode and Effects Analysis: FMEA*. Excerpts from the manual are reprinted with permission from the *FMEA Manual* (Daimler Chrysler, Ford, General Motors Supplier Quality Requirements Task Force).

Suppliers of parts and components that go into automobiles made by these three companies are required to apply the FMEA technique outlined in the reference manual. Thus, the impact of the auto industry requirement and the reference manual is immense. The criteria to be rated are severity of effect, occurrence probability, and likelihood of detection. This is how the criteria are defined (13):

- Severity of effect—is an assessment of the seriousness of the effect of the potential failure mode to the next component, subsystem, system, or customer if it occurs.
- Probability of failure—is the likelihood that a specific cause/mechanism will occur.
- Likelihood of detection—is an assessment of the ability of the proposed design controls to detect a potential cause/mechanism ...

Some professionals in risk assessment caution against having a large number of gradations within a criterion because of the inability of people to differentiate between them meaningfully when making the subjective judgments required. This FMEA system has 10 ranking levels for severity of effect and for likelihood of detection. There are five ranking levels for probability of failure, with each ranking having two possible scores. A list of the gradations follows:

CRITERIA: SEVERITY OF EFFECT	LIKELIHOOD OF DETECTION
Hazardous: without warning	Absolute uncertainty
Hazardous: with warning	Very remote
Very high	Remote
High	Very low
Moderate	Low
Low	Moderate
Very low	Moderately high
Minor	High
Very minor	Very high
None	Almost certain

PROBABILITY OF FAILURE
Very high: failure is almost inevitable
High: repeated failures
Moderate: occasional failures
Low: relatively few failures
Remote: failure is unlikely

In the AIAG reference manual, descriptions are given for each gradation to assist the raters in drawing their conclusions. Nevertheless, for severity of effect and likelihood of detection, it is difficult to establish substantive distinctions between one gradation and a higher or lower gradation.

After rankings are assigned to each of the criteria, a Risk Priority Number (RPN) is computed. It is the product of the Severity (S), Occurrence (O), and Detection (D) ranking:

$$\text{RPN} = (\text{S}) \times (\text{O}) \times (\text{D})$$

The following quotation from the AIAG text is particularly significant because of the instruction, emphasized by italics, that regardless of the RPN, special attention is to be given when severity is high. This statement relates appropriately to severity's importance in Tables 1 and 2.

> The Risk Priority Number ... is a measure of the design risk. This value should be used to rank order the concerns in the design (e.g., in Pareto fashion). The RPN will be between "1" and "1,000." For higher RPN's the team must undertake efforts to reduce this calculated risk through corrective action(s). *In general practice, regardless of the resultant RPN, special attention should be given when severity is high* (21).

After the column in which a severity ranking is entered in the Potential Failure Mode and Effects Analysis (Design FMEA) form, there is a column entitled, "Class," which stands for classification. The corresponding instruction is as follows: "Any item deemed to require special process controls should be identified on the Design FMEA form with the appropriate character or symbol in the Classification column and should be addressed in the Recommended actions column." That provides an additional control mechanism for "special product characteristics (e.g., critical, key, major, significant), components, subsystems, or systems that may require additional evaluation."

If we consider the current costs of publications and the valuable information contained within it, AIAG's FMEA Reference Manual is a buy. It costs a mere $7.

Semiconductor Manufacturing Companies' FMEAs

International SEMATECH is a consortium of 13 semiconductor manufacturing companies from 7 countries. Its *Failure Mode and Effects Analysis (FMEA): A Guide for Continuous Improvement for the Semiconductor Equipment Industry*, a 1992 document, also describes an FMEA technique having three process controls. Excerpts from that document appear here, with the permission of SEMATECH. There are similarities between this system and the one issued by the automobile manufacturing industry.

However, this system is unique in that the scope of the severity criteria is quite broad, encompassing all forms of harm, damage, or loss. Further, the value of safety engineering's counsel is recognized by including that designation—safety engineering—among the titles of people who would form a team to conduct FMEAs.

In the SEMATECH FMEA, there are three criteria to be ranked: severity, occurrence, and detection. This is how they are defined in the *Guide*. As was stated earlier in this chapter, definitions of the terms used in risk scoring systems vary, and these three have their own particular characteristics (13):

- Severity ranking criteria—Customer satisfaction is key in determining the effect of a failure mode. Safety criticality is also determined at this time based on Environmental, Safety and Health (ES&H) levels. Based on this information, a severity ranking is used to determine criticality of the failure mode on the subassembly to the end effect. The end (global) effect of the failure mode is the one to be used for determining the severity ranking. Calculating the severity levels provides for a classification ranking that encompasses safety, production continuity, scrap loss, etc.
- Occurrence ranking criteria—the probability that a failure will occur during the expected life of the system can be described in potential occurrences per unit of time.
- Detection ranking criteria—this section provides a ranking based on an assessment of the probability that the failure mode will be detected given the controls that are in place.

Two scoring tables are given in Table 3 for the severity ranking: one has five possible scorings and relates to the effect a failure may have on customer rela-

TABLE 3 Scoring Tables for Severity Ranking

Severity Ranking Criteria

Rank	Description
1–2	Failure is of such a minor nature, internal or external customer will probably not detect it.
3–5	Failure will result in slight customer annoyance and/or slight deterioration of part or system performance.
6–7	Failure will result in customer dissatisfaction and/or deterioration of part or system performance.
8–9	Failure will result in high degree of customer dissatisfaction or cause non-functionality of system.
10	Failure will result in major customer dissatisfaction and cause nonsystem operation or noncompliance with government regulations.

ES & H Severity Level Definitions

Rank	Severity Level	Description
10	Catastrophic I	A failure results in major injury or death of personnel.
7–9	Critical II	A failure results in minor injury to personnel, personnel exposure to harmful chemicals or radiation, a fire or a release of chemicals in the environment.
4–6	Major III	A failure results in a low-level exposure to personnel, or activates facility alarm system.
1–3	Minor IV	A failure results in minor system damage but does not cause injury to personnel, allow any kind of exposure to operational or service personnel, or allow any release of chemicals into environment.

SEMATECH. Failure Mode and Effects Analysis (FMEA): A Guide for Continuous Improvement for the Semiconductor Industry.

tions; the other pertains to environmental, safety, and health levels, having four possible scorings. Occurrence ranking and detection ranking criteria each have five possible scorings. Note that the versions of the criteria presented are abbreviated.

The following appears beneath the ES & H severity level definitions in the SEMATECH *Guide*:

> ES&H severity levels are patterned after the industry standard, SEMI S2-91, Product Safety Guidelines. All equipment should be designed to level IV severity. Types I, II, III are considered unacceptable risks (4.1.1).

Thus, Type IV severity defines the level of acceptability. This industry has established high environmental, safety, and health standards for itself.

A slightly condensed display of the Occurrence Ranking Criteria follows in Table 4. It interestingly relates probability to a time operating interval and expresses probability as a failure rate. The detection ranking scale of Table 5 further condenses the language of the SEMATECH *Guide* pertaining to what the verification system and/or controls are expected to accomplish.

TABLE 4 Scoring Table for Occurrence Ranking

	Occurrence Ranking Criteria
Rank	Description
1	*An unlikely probability of occurrence* during the item operating time interval. Unlikely is defined as a single failure mode (FM) probability < 0.001 of the overall probability of failure during the time operating interval.
2–3	*A remote probability of occurrence* during the item operating time interval (i.e., once every 2 months), defined as a single FM probability > 0.001 but <0.01.
4–6	*An occasional probability of occurrence* during the item operating time interval (i.e., once a month), defined as a single FM probability > 0.01 but <0.10.
7–9	*A moderate probability of occurrence* during the item time operating interval (i.e., once every 2 weeks), defined as a single FM probability > 0.10 but <0.20.
10	*A high probability of occurrence* during the item time operating interval (i.e., once a week), defined as a single FM probability > 0.20.

SEMATECH. Failure Mode and Effects Analysis (FMEA): A Guide for Continuous Improvement for the Semiconductor Industry.

As was the case in the automobile industry FMEA process, RPN is calculated:

$$\text{RPN} = \text{severity} \times \text{occurrence} \times \text{detection}$$

In terms of the decision-making process, however, there is a major difference in this system that is of particular interest to safety professionals.

The auto industry FMEA form contains a provision to enter a classification indicator for "special product characteristics (e.g., critical, key, major, significant)." In the semiconductor industry FMEA form, that column appears under the "Cr" heading and a "Y" would be entered for yes if the failure is considered safety-critical. This lends importance to the Environmental, Safety and Health

TABLE 5 Scoring Table for Detection Ranking

Rank	Detection Ranking Criteria Probability That the Defect Will Be Detected
1–2	Very high: almost certain
3–4	High: good chance
5–7	Moderate: likely
8–9	Low: not likely
10	Very low (or zero)

SEMATECH. Failure Mode and Effects Analysis (FMEA): A Guide for Continuous Improvement for the Semiconductor Industry.

Severity Level Definitions previously shown. On the form, the instruction is that critical failures must be addressed when safety is an issue.

Although the SEMATECH FMEA publication was specifically drafted for the semiconductor equipment industry, it is recommended nevertheless as a general reference. It is downloadable and free at www.sematech.org.

System Used by Some Heavy Equipment Manufacturers

Hazard analysis and risk assessment systems have been used for many years by certain heavy equipment manufacturers in their design processes to achieve inherently safer products. Keep in mind that their principle concern is product safety and the avoidance of injury to users of their equipment, or bystanders.

There are variations in the terms used in the various versions of this industry's hazard analysis system. In one version, these are the risk-assessment variables:

- Severity of the injury
- Frequency of exposure
- Probability that the event results in injury and injury is not avoided

In another version, the parameters to be scored are severity, frequency, and vulnerability. It is this latter system and its definitions that will be discussed here:

- *Severity.* The most probable injury that would be expected from an accident.
- *Frequency.* An estimate of how often a product user or bystander may be exposed to a hazard. The hazard may result from a machine part or system failure or from a man–machine interface failure.
- *Vulnerability.* The degree of user or bystander susceptibility to injury when exposed to a hazard. The likelihood that personal injury will occur once exposure to a hazard has occurred, taking into account the detectability of a hazard, risk assumptions, presence of environmental or stress conditions, skills and attitudes, etc.

For each of the three variables, there are five possible ratings, all using identical numerics: 1, 3, 5, 7, and 9. The ratings follow:

Severity Scale: S

1. Minor first aid, immediate return to work/activity
3. Doctor's office or emergency-room treatment, up to 1 week lost time
5. Hospitalization, up to 1 month lost time, no loss in work capacity
7. Permanent partial loss in work capacity, increased work difficulty
9. Death or complete disability

Frequency Scale: F

1. Theoretically can occur, but highly unlikely during the life of the machine population
3. Only once in the life of a small percentage (10%) of the product
5. Once per use season or once annually
7. Once daily
9. Continuous exposure

Vulnerability Scale: V

1. Practically impossible to complete injury sequence
3. Remotely possible, but unlikely
5. Some conditions favorable to completing the injury sequence
7. Very possible, but not assured
9. Almost certain to complete injury sequence

$$\text{Risk Score} = \text{Severity} \times \text{Frequency} \times \text{Vulnerability.}$$

The literature on this system indicates the following:

- Hazards analyses and risk assessments are completed by a team.
- Consensus is to be reached on risk scores.
- Risk scores are listed and ranked.
- High and low values would be examined, and actions for improvement developed when required.

This risk scoring system gives equal weight to all variables. Thus, the needed emphasis on severity of injury is diminished when the risk score is produced. In a follow-up discussion with one user of this system, it was acknowledged that, with all variables being given equal weight, the significance of severity became subordinate. However, it was also said that, in practice, the team gave the potential for severe injury due consideration when developing recommendations for product improvement.

The William T. Fine System

While at the Naval Ordnance Laboratory in White Oak, Maryland in 1971, William T. Fine submitted a report entitled, "Mathematical Evaluations for Controlling Hazards." In its abstract, Fine made the following comment about his work:

> ... A formula has been devised which weighs the controlling factors and "calculates the risk" of a hazardous situation, giving a numerical evaluation to the urgency for remedial attention to the hazard. Calculated Risk Scores are then used to establish priorities for corrective action. An additional formula weighs the estimated cost and effectiveness of any contemplated corrective action against the Risk Score and gives a determination as to whether the cost is justified.

Fine's system was well publicized. Variations of it remain in use. In light of the later emergence of risk scoring systems, his report is interesting and thought-provoking. This is Fine's risk score formula:

$$\text{Risk Score} = \text{Consequences} \times \text{Exposure} \times \text{Probability}$$

$$\text{Abbreviated: R} = \text{C} \times \text{E} \times \text{P}$$

His definition of consequences is comparable to that appearing in other risk scoring systems. However, his definitions of exposure and probability are unique:

- Consequences: The most probable results of a potential accident, including injuries and property damage.
- Exposure: Frequency of occurrence of the hazard-event (the undesired event which could start the accident-sequence).
- Probability: This is the likelihood that, once the hazard-event occurs, the complete accident-sequence of events will follow with the necessary timing and coincidence to result in the accident and consequences.
- Hazard-event: An undesirable occurrence; the combination of a hazard with some activity or person which could start a sequence of events to end in an accident.
- Accident sequence: The chain of events or occurrences which takes place starting with a "hazard-event" and ending with the consequences of an accident.

Fine's use and definition of the term "accident-sequence" relate closely to what Hendrick and Benner wrote about the beginning and end of an accident in *Investigating Accidents With STEP* (STEP means sequentially timed events plotting):

> Because accidents are composed of sets of individual events, all of which are interrelated, each event affects one or more actors and what they do next, changing their state. The first event in the accident process is a perturbation or an undesired

> or unplanned change by someone or something within the planned process. That first disruptive event initiates a sort of cascading effect, culminating in some harm or loss (31). . . .
>
> The term "actor" is used to identify the people or things who or which directly influenced the flow of events constructing the accident sequence (69).

Fine's report indicates that the values assigned to the various factors are empirical and derive from his best judgment. His report was submitted in 1971. That should be kept in mind when considering estimates and/or costs.

TABLE 6

Consequences: Most Probable Result of the Potential Accident

Classification	Rating
a. Catastrophe, numerous fatalities, damage over $1,000,000; major disruption of activities	100
b. Multiple fatalities; damage over $500,000 to $1,000,000	50
c. Fatality, damage $100,000 to $500,000	25
d. Extremely serious injury (amputation, permanent disability); damage $1,000 to $100,000	15
e. Disabling injury; damage up to $1,000	5
f. Minor cuts, bruises, bumps; minor damage	1

Exposure: The Frequency of Occurrence of the Hazardous Event

Hazard Event Occurs	Rating
a. Continuously, or many times daily	10
b. Frequently, approximately once daily	6
c. Occasionally, from once per week to once per month	3
d. Unusually, from once per month to once per year	2
e. Rarely, but it has been known to occur	1
f. Remotely possible, but not known to have occurred	0.5

Probability: Likelihood That Accident Sequence Will Follow to Completion

Complete Accident Sequence	Rating
a. Is the most likely and expected result if the hazard event takes place	10
b. Is quite possible, not unusual, has an even 50/50 chance	6
c. Would be an unusual sequence or coincidence	3
d. Would be a remotely possible coincidence	1
e. Has never happened after many years of exposure, but is conceivably possible	0.5
f. Practically impossible sequence, has never happened	0.1

Fine's rating system as outlined in Table 6 gives appropriate weightings to consequences (the severity of harm or damage) in relation to the exposure and probability categories, for which the highest rating possible is 10. To illustrate: Numerous fatalities are given a rating of 100; multiple fatalities a rating of 50; a single fatality a rating of 25. Thus, although Fine's system is three-dimensional, it does not result in a diminished status for severity.

TABLE 7 Hazard Descriptions

Action level: Immediate correction required. Activity should be discontinued until hazard is reduced.

Hazard Description	Risk Score
Window washer on third floor, without safety safety belt, hangs on with one hand and leans out	1500
Men smoking in flammable storage warehouse, sprinkler system, highly flammable material	270

Action level: Urgent. Requires attention as soon as possible.

Hazard Description	Risk Score
Portable electric drill in use without ground wire, getting rough usage by several people	200
Steps of main building slippery whenever wet, no handrail, many pedestrians daily	90

Action level: Hazard should be eliminated without delay, but situation is not an emergency.

Hazard Description	Risk Score
Compressed oxygen cylinder standing unsecured near wall, little traffic or movement	85
Workman using hammer with loose head, in use daily for odd jobs	18

Although Fine does not specifically list categories of risk scores and the recommended management action levels for them, three action levels are implied in an illustrative listing of 25 hazard descriptions. See Table 7. For each management action level, the first and last hazard description in his list, and its Risk Score, are presented. Note that for the highest scores, the first action level, operation is not permissible.

Fine also produced a formula to determine whether "the cost of corrective action to alleviate a hazard is justified." In applying his formula for Justification (J), the Risk Score is divided by a Cost Factor (CF) multiplied by a Degree of Correction Factor (DC). The formula, and the ranges for the ratings of the Cost Factors and the degree-of-correction factors, are given in Table 8. Keep in mind that the numbers are subjectively determined.

$$J = \frac{C \times E \times P}{CF \times DC}$$

The Critical Justification Rating, as Fine states, is arbitrarily set at 10. For any rating over 10, the expenditure will be considered justified. For a score of less than 10, the cost of contemplated corrective action is not justified.

Fine's system is thought-provoking. Sometimes, application of it leads to the observation that relying entirely on a numerical approach to determining

TABLE 8

Cost Factors

Cost	Rating
Over $50,000	10
$25,000 to 50,000	6
$10,000 to 25,000	4
$1,000 to 10,000	3
$100 to 1,000	2
$25 to 100	1
Under $100	0.5

Degree of Correction Factors

Description	Rating
Hazard positively eliminated, 100%	1
Hazard reduced at least 75%, but not completely	2
Hazard reduced by 50–75%	3
Hazard reduced by 25–50%	4
Slight effect on hazard, less than 25%	6

whether risks are to be ameliorated produces questionable conclusions. Consider these applications of his justification method. Assume the following:

	Rating
Consequence: fatality	25
Exposure: from once per month to once per year	2
Probability: has an even 50/50 chance	6

$$\text{Risk score} = 25 \times 2 \times 6 = 300$$

Amelioration cost:
  Over $50,000—CF factor is 10
  $25,000 to 50,000—CF factor is 6
Degree of correction:
  At least 75%—DC factor is 2
  25–50%—DC factor is 4

$$300 \div (10 \times 2) = 15 \qquad 300 \div (10 \times 4) = 7.5$$

$$300 \div (6 \times 2) = 25 \qquad 300 \div (6 \times 4) = 12.5$$

The Justification level was arbitrarily set at 10. For a fatality with an even chance of occurrence, in one of these four scenarios, the expenditure would not be considered justified. This is the sort of situation that requires executive review. If the expenditure is not to be made, alternate methods of risk reduction must be applied. This system, overall, could produce some interesting discussions.

The Kinney and Wiruth Innovation

In 1976, while at the Naval Weapons Center in China Lake, California, G. F. Kinney and A. D. Wiruth developed a report entitled, "Practical Risk Analysis for Safety Management." The authors reference Fine's earlier work and state that many of their evaluations are based on his report.

Kinney and Wiruth also produce a risk score, considering likelihood, exposure, and consequences, but the definitions of terms used differ from Fine's. In addition, the numerical values given for the classifications within each category differ somewhat. Their numerical values are not repeated here.

What Kinney and Wiruth did that was somewhat different was to produce two nomographs: one for Risk Analysis; the other for Cost Effectiveness Factors. Both are presented here in Figures 1 and 2. These nomograms have appeared in several texts and have been the subject of a number of talks and discussions.

For the Risk Analysis nomogram, one would first determine the classifications to be given within the likelihood, exposure, and consequence categories. Then, Kinney and Wiruth's instruction applies on how to use the Risk Analysis nomograph:

> To calculate a risk score using this nomograph, locations corresponding to each factor involved are first established. Then a line is drawn from the point for the likelihood factor through that for the exposure factor and extended to the tie line in the center. (Location along this tie line corresponds to the product of these two factors but the numbers have been omitted.) A second line is drawn from this point on the tie line through that for the consequence factor and extended for the risk score. A numerical value for this risk score and its descriptive equivalent are then obtained directly (10).

Five action levels appear on the right side of the nomogram. Their descriptions are as follows:

Risk Score	Risk Situation
>400	Very high risk; consider discontinuing operation
200–400	High risk; immediate correction required
70–200	Substantial risk; correction needed
20–70	Possible risk; attention indicated
<20	Risk; perhaps acceptable

Since the numerical values would have been determined for likelihood, exposure, and consequence, the risk score could have been developed by simple multiplication and then related to action levels. Nonetheless, supporting graphics can be impressive.

Despite their widespread use, some of the risk levels and risk management indicators that derive from Kinney and Wiruth's Risk Analysis nomograph would be considered by many to be inappropriate. For the exercise next sug-

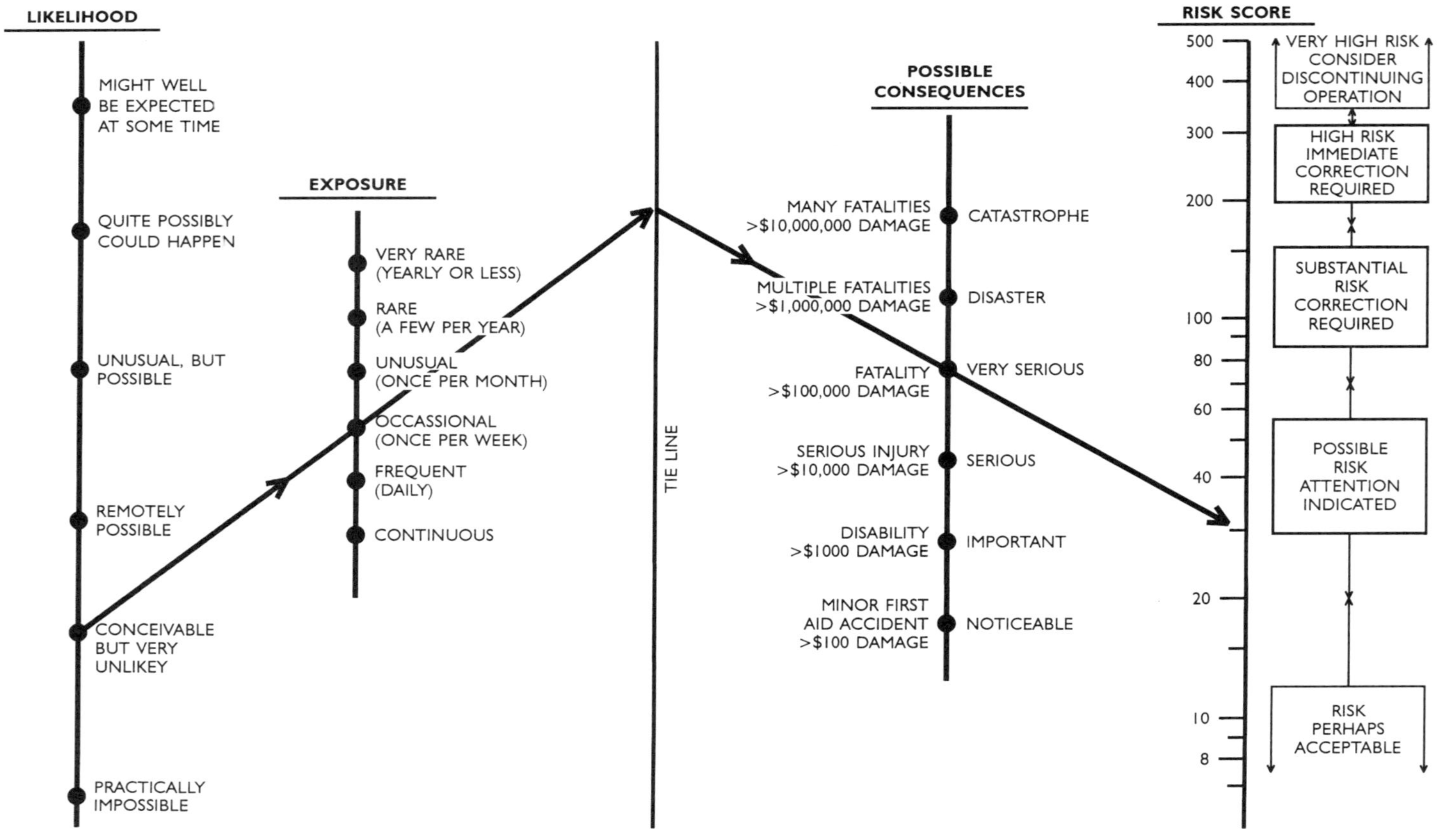

FIGURE 1 *Risk analysis.*

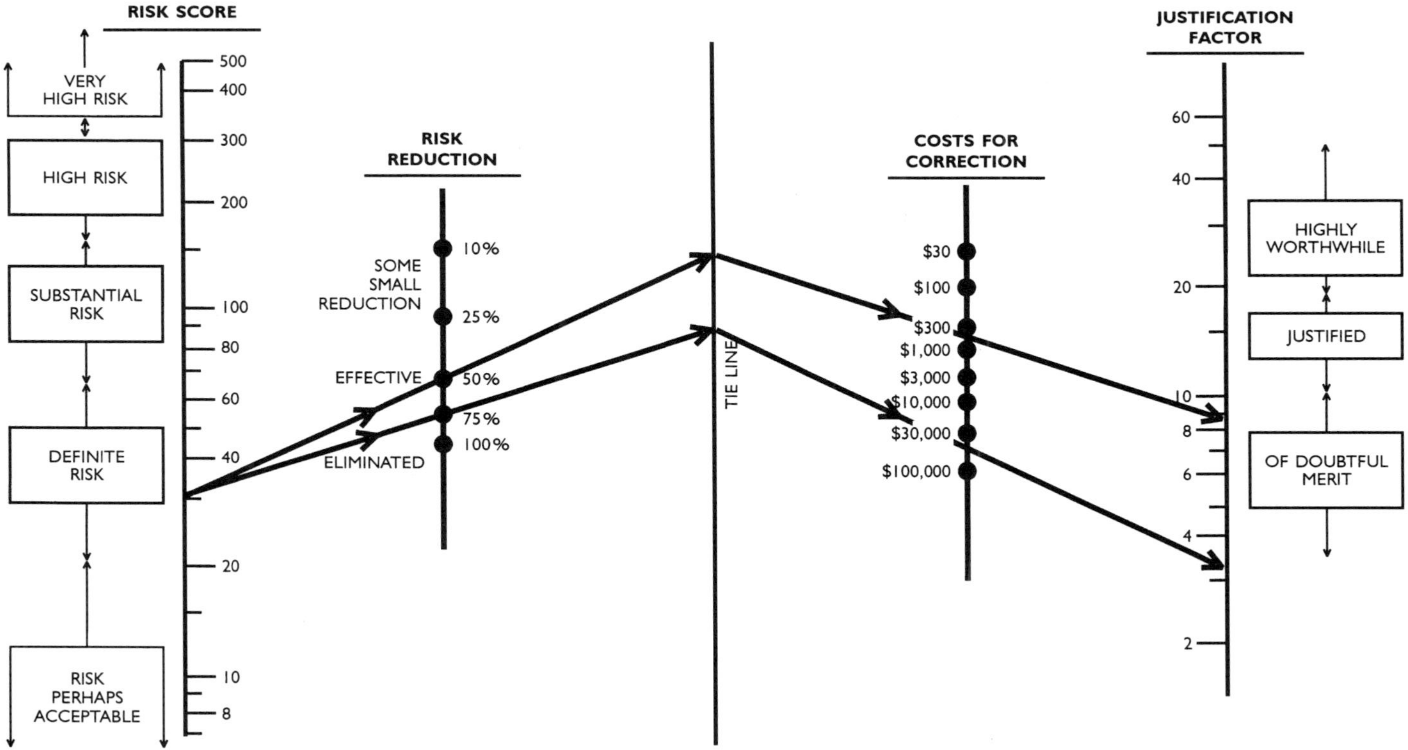

FIGURE 2 *Cost-effectiveness factors.*

gested, an understanding would be more easily attained if one were available to draw lines on copies of the nomograph. In the Risk Analysis nomograph:

- Draw a line from the Likelihood category "remotely possible" through the Exposure category "rare—a few per year" to the Tie Line.
- From that point on the Tie Line, draw a line through "Disaster" in the Possible Consequence column to the Risk Score line.
- Opposite that point on the Risk Score line, the reading is "Possible Risk Reduction Indicated."

Of the five management action categories shown, this management action Possible Risk Reduction Indicated is the option second from the bottom and inappropriate for a Disaster. Start again, on the Risk Analysis nomograph:

- Draw a line from the Likelihood category "Remotely Possible" through the Exposure category "Rare—a few per year" to the Tie Line.
- From that point, draw a line through the Possible Consequence category "Very Serious—Fatality" to the Risk Score line.
- The indication is "Risk Perhaps Acceptable."

This author believes it inappropriate to suggest such low risk categories when fatalities might occur in the circumstances just previously described.

To continue and determine the justification of the cost for risk amelioration, let us use the cost-effectiveness nomograph of Figure 2:

- The risk score previously determined is located on the Risk Score scale shown.
- A line is drawn through the point representing the percent of Risk Reduction to be obtained to the Tie Line.
- From there, a line is drawn through the appropriate point in the Costs for Correction column to the right-hand scale showing Justification Factors.

As is the case in the Fine system, a justification value of less than 10 indicates that the risk reduction proposal is of doubtful merit. Three justification levels are given: Highly Worthwhile, Justified, and Of Doubtful Value.

Kinney and Wiruth do not suggest an arbitrary use of cost-effectiveness determinations. Their report contains an example of a justification calculation, about which they make the following observation:

> A cost effectiveness of only 3.3 indicates that the suggestion is of doubtful merit. But since the risk score of 30 indicates that some possible risk is present, it appears that alternative methods for risk control should be investigated (15).

This brings the use of their system a little closer to the intent of Tables 1 and 2—just a bit.

A FOUR-DIMENSIONAL RISK SCORING SYSTEM

Although reference will be made here to the four-dimensional risk scoring system as it appears in *Pilz: Guide to Machinery Safety*, 6th edition, it should be understood that this system has appeared elsewhere. For example, it has been used, with minor modification, by at least one company to meet the requirements of the European Standard EN 1050, "Safety of Machinery—Principles for risk assessment."

In the Pilz text, early in Chapter 4 entitled, "Background to risk assessment," the following statement is made:

> In simple terms, there are only 2 real factors to consider:
>
> - The severity of foreseeable injuries (ranging from a bruise to a fatality)
> - The probability of their occurrence.

As the text proceeds, the system becomes more complex. There are four elements to be scored, not two:

- Likelihood of occurrence/contact with hazard (LO)
- Frequency of exposure to the hazard (FE)
- Degree of possible harm (DPH), taking into account the worst possible case
- Number of persons exposed to the hazard (NP)

This system has a particular focus: It applies to personal injury that could derive from machinery operation. As will be observed, all the listings for "Degree of possible harm" involve personal injuries. There are no entries for possible damage to property or the environment.

Terms used to establish gradations and scores in the Likelihood of occurrence and Frequency of exposure categories are comparable to those in previously cited risk assessment systems. They follow in the next few lists here. Note the place given, relatively, within the listings to "Even chance—could happen" in the Likelihood category and "Annually" in the Frequency of exposure category. A computation appears later in which these ratings are used.

LIKELIHOOD OF OCCURRENCE (LO) AND THE SCORES

Almost impossible: possible only under extreme circumstances	0.033
Highly unlikely: although conceivable	1
Unlikely: but could occur	1.5
Possible: but unusual	2
Even chance: could happen	5
Probable: not surprising	8
Likely: only to be expected	10
Certain: no doubt	15

FREQUENCY OF EXPOSURE TO THE HAZARD (FE) AND THE SCORES

Annually	0.5
Monthly	1
Weekly	1.5
Daily	2.5
Hourly	4
Constantly	5

As has already been said, the listing for degree of possible harm enumerates specific injury types, and it is somewhat unusual in that respect.

DEGREE OF POSSIBLE HARM AND THE SCORES

Scratch/bruise	0.1
Laceration/mild ill-effect	0.5
Break minor bone or minor illness (temporary)	2.0
Break major bone or major illness (temporary)	4.0
Loss of one limb, eye, hearing loss (permanent)	6.0
Loss of two limbs, eyes (permanent)	10.0
Fatality	15.0

The fourth category in this system is Number of persons exposed to the hazard:

NUMBER OF PERSONS EXPOSED AND THE SCORES

1–2 persons	1
3–7 persons	2
8–15 persons	4
16–50 persons	8
50+ persons	12

Using the numerical ratings given to each element, the formula produces a risk level by simple multiplication. The formula and risk levels follow:

$$\text{LO} \times \text{FE} \times \text{DPH} \times \text{NP} = \text{Risk level}$$

RISK LEVEL	SCORING RANGE
Negligible: presenting very little risk to health and safety	0–5
Low but significant: containing hazards that require control measures	5–50
High: having potentially dangerous hazards, which require control measures to be implemented urgently	50–500
Unacceptable: continued operation in this state is unacceptable	500+

TABLE 9 Usage of Four-Dimensional Risk-Scoring System

Ratings	Number of Persons Exposed				
	2	7	15	50	51
Likelihood: even chance	5	5	5	5	5
Frequency of exposure: annual	0.5	0.5	0.5	0.5	0.5
Degree of possible harm: fatality	15	15	15	15	15
For: number of persons exposed	1	2	4	8	12
Risk levels	37.5	75	150	300	450
	Low but significant	High	High	High	High

Assume that in a company's annual shutdown for retooling and maintenance, a singular task was to be performed for which the likelihood of occurrence of a hazardous event was rated as "even chance—could happen" and the outcome would be fatalities. In the example shown in Table 9, the ratings are the same for likelihood of occurrence, frequency of exposure, and degree of possible harm. Ratings vary for the number of people exposed.

For such a hazard scenario, the computations never fall within the Unacceptable risk level (500+), even when there is an even chance that 51 people will be killed. Score levels in this system thus seem ill-conceived. They greatly diminish the value of life. Nevertheless, this risk scoring system continues to be promoted.

The use of the number of persons exposed as a category in risk assessment requires careful consideration. One fatality would be considered a catastrophe in Table 1. In the FMEA system published by SEMATECH, a fatality is also a catastrophe. In the Fine system, a fatality that has an even chance of occurring and an annual exposure would fall in the unacceptable risk category, for which operations are to be discontinued until the hazard is reduced. Not so in this system, either for a single fatality or 51 fatalities. That is simply not acceptable.

MODEL THREE-DIMENSIONAL RISK SCORING SYSTEM

One of our aims in making this study of multidimensional risk scoring systems was to determine whether a model could be proposed that:

- Serves the needs of those who are more comfortable with statistics in their risk assessments
- Addresses the strong beliefs of those who want frequency of exposure given separate consideration in the risk assessment process (e.g., having frequency of exposure included as an element in a risk decision matrix was much debated in the meetings out of which the B11.TR3 report was developed)
- Maintains credibility and efficacy

As the work proceeded in crafting the risk scoring model presented here, these conclusions were drawn.

- In a statistical risk scoring system, all scores have to meet a plausibility test. High-risk scores must be produced for high risks; a lower-level risk should not fall in a high-risk category.
- To create a focal point for the relative placing of other risk scores, it was assumed that for an incident having a severity of one or more fatalities, with an even chance of occurring (50/50), and an annual frequency of exposure, the risk score would be high.
- Frequency of exposure can be separately evaluated without negatively affecting the validity of risk determinations, provided that adequate weighting is given to severity of outcome in the scoring system. (Some professionals in risk assessment argue that frequency of exposure is one of the factors considered in determining probability and ought not to be separately identified.)
- Adaptations can be made of the risk assessment matrices shown in Chapter 6, "A Primer on Hazard Analysis and Risk Assessment," to develop a single risk scoring model that addresses the following: injury to people (employees, the public); facilities, product, or equipment loss; operations down time; chemical releases and environmental damage.
- If the number of gradations for probability, frequency of exposure, or severity is excessive, distinctions between them are difficult to make: Five gradations were chosen for probability and for frequency of exposure; for severity of outcome, there are four.
- Although an attempt was made to not have the same term appear in elements to be scored, it was unsuccessful.
- A three-dimensional risk scoring system is plausible.

Definitions follow of the terms relevant to our risk scoring system:

- *Hazard.* The potential for harm; hazards include the characteristics of things and the actions or inactions of people; hazards encompass all aspects of technology or activity that produce risk.
- *Probability.* An expression of the likelihood of the potential of a hazard being realized and initiating an event or series of events that could result in harm or damage—for the selected unit of time, events, population, items, or activity.
- *Frequency of Exposure.* An estimate of the frequency and duration of exposure to the hazard, over time; frequency of endangerment, over time.
- *Severity.* An estimate of the worst credible consequences, should a hazard-related incident occur.
- *Risk.* A measure of the probability of a hazard-related incident occurring and the severity of harm or damage that could result.

The Risk Score Formula

Once we decided to give frequency of exposure separate consideration as an element to be scored, the question became how to assure that such an element did not skew the scoring computation. All elements need not be multipliers, as is the case in the systems previously described.

In this Risk Score Formula, the rating for frequency of exposure is added to the rating for occurrence probability to produce a mid-level score that is then multiplied by the severity rating:

$$\text{Risk Score} = (\text{Probability Rating} + \text{Frequency of Exposure Rating}) \times \text{Severity Rating}$$

$$\text{RS} = (\text{PR} + \text{FER}) \times \text{SR}$$

Gradation and Scoring Development

To maintain plausibility in accord with the content of Tables 1 and 2, a variety of scorings were tested until credible results were obtained. The gradations and ratings selected are shown in Table 10.

TABLE 10 Descriptive Words and Ratings

Probability		Frequency of Exposure		Severity	
Frequent (Fre)	15	Often (Of)	13	Catastrophic (Cat)	50
Likely (Lik)	9	Occasional (Oc)	10	Critical (Cri)	40
Occasional (Occ)	4	Infrequent (In)	7	Medium (Med)	25
Remote (Rem)	1	Seldom (Se)	4	Minimal (Min)	10
Improbable (Imp)	0.5				

Risk Categories, Score Levels, and Action Levels

The Risk Categories in Table 11 are identical to those shown in Table 2, as are the indicators for Remedial Action, Or Acceptance. Score levels have been added. It must be understood that score levels are advisory, having been established from subjective judgments. They should not be considered as absolutes in management decision making. For example, in a real-world case having a high

TABLE 11 Risk Categories, Score Levels, and Action Levels

Risk Category	Score Levels	Remedial Action, or Acceptance
High	800 and above	Operation not permissible
Serious	500–799	Remedial actions to have high priority
Moderate	200–499	Remedial actions to be taken in appropriate time
Low	199 and below	Risk is acceptable: remedial action discretionary

severity potential, the remedial actions taken brought down the risk score to 225, which is very close to 199, the level where the risk would be considered acceptable. Understandably, the work involved proceeded, with close observation.

Plausibility and Efficacy Test

All combinations of occurrence probability, frequency of exposure, and severity of outcome appear in Tables 12 through 16. The purpose of these tables is to establish the plausibility and efficacy of the risk scoring system. Table 16 is a composite of Tables 12 through 15.

TABLE 12 Plausibility and Efficacy Test 1: Probability, Frequency of Exposure, and Catastrophic Severity

Probability Category and Rating		Frequency of Exposure and Rating		Mid-score	Catastrophic Severity	Final Score
Fre	15	Of	13	28	50	1,400
	15	Oc	10	25	50	1,250
	15	In	7	22	50	1,100
	15	Se	4	19	50	950
Lik	9	Of	13	22	50	1,100
	9	Oc	10	19	50	950
	9	In	7	16	50	800
	9	Se	4	13	50	650
Occ	4	Of	13	17	50	850
	4	Oc	10	14	50	700
	4	In	7	11	50	550
	4	Se	4	8	50	400
Rem	1	Of	13	14	50	700
	1	Oc	10	11	50	550
	1	In	7	8	50	400
	1	Se	4	5	50	250
Imp	0.5	Of	13	13.5	50	675
	0.5	Oc	10	10.5	50	525
	0.5	In	7	7.5	50	375
	0.5	Se	4	4.5	50	225

TABLE 13 Plausibility and Efficacy Test 2: Probability, Frequency of Exposure, and Critical Severity

Probability Category and Rating		Frequency of Exposure and Rating		Mid-score	Critical Severity	Final Score
Fre	15	Of	13	28	40	1,120
	15	Oc	10	25	40	1,000
	15	In	7	22	40	880
	15	Se	4	19	40	760
Lik	9	Of	13	22	40	880
	9	Oc	10	19	40	760
	9	In	7	16	40	640
	9	Se	4	13	40	520
Occ	4	Of	13	17	40	680
	4	Oc	10	14	40	560
	4	In	7	11	40	440
	4	Se	4	8	40	320
Rem	1	Of	13	14	40	560
	1	Oc	10	11	40	440
	1	In	7	8	40	320
	1	Se	4	5	40	200
Imp	0.5	Of	13	13.5	40	540
	0.5	Oc	10	10.5	40	420
	0.5	In	7	7.5	40	300
	0.5	Se	4	4.5	40	180

TABLE 14 Plausibility and Efficacy Test 3: Probability, Frequency of Exposure, and Medium Severity

Probability Category and Rating		Frequency of Exposure and Rating		Mid-score	Medium Severity	Final Score
Fre	15	Of	13	28	25	700
	15	Oc	10	25	25	625
	15	In	7	22	25	550
	15	Se	4	19	25	475
Lik	9	Of	13	22	25	550
	9	Oc	10	19	25	475
	9	In	7	16	25	400
	9	Se	4	13	25	325
Occ	4	Of	13	17	25	425
	4	Oc	10	14	25	350
	4	In	7	11	25	275
	4	Se	4	8	25	200
Rem	1	Of	13	14	25	350
	1	Oc	10	11	25	275
	1	In	7	8	25	200
	1	Se	4	5	25	125
Imp	0.5	Of	13	13.5	25	338
	0.5	Oc	10	10.5	25	263
	0.5	In	7	7.5	25	188
	0.5	Se	4	4.5	25	113

TABLE 15 Plausibility and Efficacy Test 4: Probability, Frequency of Exposure, and Minimal Severity

Probability Category and Rating		Frequency of Exposure and Rating		Mid-score	Minimal Severity	Final Score
Fre	15	Of	13	28	10	280
	15	Oc	10	25	10	250
	15	In	7	22	10	220
	15	Se	4	19	10	190
Lik	9	Of	13	22	10	220
	9	Oc	10	19	10	190
	9	In	7	16	10	160
	9	Se	4	13	10	130
Occ	4	Of	13	17	10	170
	4	Oc	10	14	10	140
	4	In	7	11	10	110
	4	Se	4	8	10	80
Rem	1	Of	13	14	10	140
	1	Oc	10	11	10	110
	1	In	7	8	10	80
	1	Se	4	5	10	50
Imp	0.5	Of	13	13.5	10	135
	0.5	Oc	10	10.5	10	105
	0.5	In	7	7.5	10	75
	0.5	Se	4	4.5	10	45

TABLE 16 Plausibility and Efficacy Test Summary

Probability Category and Rating		Frequency of Exposure and Rating		Mid-score	Severity Categories, Ratings, and Final Risk Scores								
					Catastrophic	Final Score	Critical	Final Score	Medium	Final Score	Minimal	Final Score
Fre	15	Of	13	28	50	1,400	40	1,120	25	700	10	280
	15	Oc	10	25	50	1,250	40	1,000	25	625	10	250
	15	In	7	22	50	1,100	40	880	25	550	10	220
	15	Se	4	19	50	950	40	760	25	475	10	190
Lik	9	Of	13	22	50	1,100	40	880	25	550	10	220
	9	Oc	10	19	50	950	40	760	25	475	10	190
	9	In	7	16	50	800	40	640	25	400	10	160
	9	Se	4	13	50	650	40	520	25	325	10	130
Occ	4	Of	13	17	50	850	40	680	25	425	10	170
	4	Oc	10	14	50	700	40	560	25	350	10	140
	4	In	7	11	50	550	40	440	25	275	10	110
	4	Se	4	8	50	400	40	320	25	200	10	80
Rem	1	Of	13	14	50	700	40	560	25	350	10	140
	1	Oc	10	11	50	550	40	440	25	275	10	110
	1	In	7	8	50	400	40	320	25	200	10	80
	1	Se	4	5	50	250	40	200	25	125	10	50
Imp	0.5	Of	13	13.5	50	675	40	540	25	338	10	135
	0.5	Oc	10	10.5	50	525	40	420	25	263	10	105
	0.5	In	7	7.5	50	375	40	300	25	188	10	75
	0.5	Se	4	4.5	50	225	40	180	25	113	10	45

What the Descriptive Words Mean

To give substance to the words used to establish gradations within the probability, frequency of exposure, and severity categories, their meanings, as they are used in this risk scoring system, are now presented in Tables 17 through 19. The risk assessment matrices shown in Chapter 6 provide an excellent reference for them.

TABLE 17 Incident Probability

Category: Descriptive Word	Definition: Applies to the Selected Unit of Time, Events, Population, Items, or Activity
Frequent	Likely to occur repeatedly, to even chance
Likely	Likely to occur several times
Occasional	Occurs sporadically, likely to occur sometime
Remote	Not likely to occur, but could possibly occur
Improbable	So unlikely, can assume occurrence will not be experienced

TABLE 18 Frequency of Exposure

Category: Descriptive Word	Definition
Often	Continuous to daily
Occasional	Daily to monthly
Infrequent	Monthly to yearly
Seldom	Less than yearly

TABLE 19 Severity of Consequences

Category: Descriptive Word	People: Employees, Public	Facilities, Product, or Equipment Loss	Operations Down time	Environmental Damage
Catastrophic	One or more fatalities	Exceeds $2M	Exceeds 4 months	Major event, requiring several years for recovery
Critical	Disabling injury or illness	500K to $2M	2 weeks to 4 months	Event requires 1 to 5 years for recovery
Medium	Minor injury, lost work day	5K to 500K	1 day to 2 weeks	Recovery time is less than 1 year
Minimal	First-aid case, minor medical treatment	Less than 5K	Less than 1 day	Minor damage, easily repaired

CONCLUSION

Thus, a three-dimensional risk scoring system can be crafted, one with credibility. But, how is such a system to be used? Risk scores produced can influence decision making and priority-setting, but in reality they are not the sole or absolute determinant.

In gathering data for this chapter, one of the most interesting discussions of the real-world use of a risk scoring system took place with a person whose principle interest was the safety of the products her company produced. The system in place required the use of an independent facilitator and a review committee of five to seven knowledgeable people. Consensus on risk scores must be reached, and recommendations made on any necessary actions.

And, this is exactly where other factors come into the discussions. Recommendations are often influenced by possible damage to one's corporate image, customer relations, and an assumed societal responsibility. In such deliberations, it is common for injuries to users or bystanders to be moved up a level or two in the severity category.

For example, an injury requiring a visit to a doctor but involving an immediate return to work receives the lowest severity rating in the scoring system. However, in concluding on recommendations to be made affecting product design or operating procedures as established in instruction manuals, a review group regularly gives more weight to an injury that the scoring system indicates is necessary.

Risk scoring systems can serve a real need. But, it should be remembered that they provide numerics on subjective judgments. Risk assessment is still as much art as science.

Several references have been made in this chapter to William T. Fine's paper, "Mathematical Evaluations for Controlling Hazards." It is thought-provoking, valuable, and somewhat hard to find. The appendix to this book is an abridged version of it. The substance of the original paper has been maintained.

REFERENCES

Accident Prevention Manual for Business and Industry: Administration and Programs, 11th ed. Itasca, IL: National Safety Council, 1997.

"Risk Assessment and Risk Reduction—A Guideline to Estimate, Evaluate and Reduce Risks Associated with Machine Tools." ANSI Tech. Rept. B11.TR3., 2000.

Failure Mode and Effects Analysis (FMEA): A Guide for Continuous Improvement for the Semiconductor Equipment Industry. Technology Transfer #92020963A-ENG. Austin, TX: International SEMATECH, 1992.

Fine, William T. "Mathematical Evaluations for Controlling Hazards." White Oak, MD: Naval Ordinance Laboratory, 1971.

Grose, Vernon L. *Managing Risk: Systematic Loss Prevention for Executives*. Englewood Cliffs, NJ: Prentice-Hall, 1987 (now available through Omega Systems Group, Incorporated, Arlington, VA).

Hendrick, Kingsley and Ludwig Benner Jr. *Investigating Accidents With STEP*. New York: Marcel Dekker, 1987.

Kinney, G. F. and A. D. Wiruth. "Practical Risk Analysis for Safety Management." China Lake, CA: Naval Weapons Center, NWC Tech. Publ. 5865, 1976.

Guide to Machinery Safety, 6th ed. Farmington, MI: Pilz Automation Technology, Feb. 1999.

Potential Failure Mode and Effects Analysis: FMEA. Southfield, MI: Automotive Industry Action Group (AIAG), 1995.

"Safety of Machinery—Principles for Risk Assessment." European Standard EN 1050, Brussels: Central Secretariat, European Committee for Standardization, 1997.

10

Current Developments in Behavioral Safety

INTRODUCTION

Unfortunately, too many speakers and writers have created the impression that behavioral safety is the panacea, by itself, that will solve all safety problems. They do the world a disservice. However, there have been significant developments in behavioral safety, and a significant change in the rhetoric has thus taken place.

Recent speeches and papers presented by representatives of two prominent consulting organizations indicate that there are conceptual transitions in progress that significantly redefine what has been said and written previously about behavioral safety. These organizations are Behavioral Science Technology, for which Thomas R. Krause is Chief Executive Officer, and Safety Performance Solutions, for which E. Scott Geller is a senior partner and founding member.

What exactly is happening in the field of behavior safety? Some of the key players now seem to:

- Accept that behavioral safety is but one element in an overall safety initiative.
- Have abandoned the oft-stated premise that 85 to 95% of accidents are principally caused by the unsafe acts of workers.
- Recognize that for most hazards-related incidents there are multiple causal factors.
- Agree that analyses of the data collected on at-risk behavior, so-called, must be made to determine the diverse sources of the causal factors.

- Accept that application of the hierarchy of controls is the most effective means of reducing risk.
- Promote the principle proposed by W. Edwards Deming that "performance does not come from the individual ... but from the system."
- Recognize that the principle focus for safety improvement should not be on the psychology of correcting worker behavior; rather the focus should be on the design of the workplace, the work methods, and management systems.

These changes in concept address many of the points raised by those who have found behavioral safety as it has been presented difficult to accept, and they will serve to reduce the confusion about what behavioral safety is and is not. Is there confusion? That is attested to by the writings of some of the key players in behavioral safety. Also, at least a few of them publicly question the substance of the work other consultants are doing, which prompts reference to the behavioral safety symposium held by the American Society of Safety Engineers (ASSE) in February 1998.

Many of the prominent consultants in the field were presenters at that conference. Their speeches, representing a wide diversity of views, are contained in *Proceedings: A Behavioral Safety Symposium* published by ASSE. That publication is recommended for an education that explores the breadth of concepts on which the behaviorists base their methods. Reading it, one will gain an appreciation of why considerable confusion has arisen about the variations in the practices of behavioral safety consultants.

This chapter will explore the confusion. Then, the proper role for behavioral safety within an overall safety initiative will be discussed. Comments will follow on the transitions occurring in behavioral safety that will eliminate some of the confusion.

WHY THE CONFUSION?: THE TWO DIVERSE BEHAVIORAL SAFETY MODELS

According to E. Scott Geller, who is a frequent writer and speaker on behavioral safety, a problem of consequence has arisen because some of those who offer themselves as behavioral consultants do not, in his view, really practice behavior-based safety. Geller, who is a practitioner in worker-focused, behavior-based safety, wrote an article entitled, "Confusion, Controversy, and Clarification." The following excerpt from that article pertains to the February 1998 ASSE conference on behavioral safety:

> This ASSE conference was designed in part to clarify the principles and methods of BBS.... This mission was not accomplished. In fact, many presentations at this conference caused more confusion.... Part of the discord and confusion about BBS ... occurred because several presenters do not practice BBS in their consulting, and thus, did not teach lessons consistent with BBS principles (40).

Note that Geller said "several presenters do not practice BBS in their consulting." Who practices behavioral-based safety and who does not, and what are their qualifications? Recent events suggest that it is important for safety practitioners to understand the major differences between the two diverse approaches being taken by consultants in behavioral safety.

One school advances a *culture change model*; the other advocates a *worker-focused behavior modification model.* Many safety practitioners have experienced the confusion brought about by situations like the one described in the following example.

In August 1998, the author wrote an article entitled, "Perspectives on ASSE's Behavioral Safety Symposium." It was published in *Professional Safety*, the magazine of the ASSE (32). The article explored, among other issues, the major differences in behavioral safety concepts offered by presenters at the February 1998 symposium.

One of the requests made for permission to reprint the article came from the safety director at a large company who wanted to reprint 100 copies. He revealed during a follow-up phone call his intention to distribute and discuss it at a gathering of all company safety personnel. In that company, purchasing authority at a location level permitted safety practitioners to purchase the behavioral safety consulting services they wanted. And, they bought services from five different consultants, all of them prominent and well known. When the location safety practitioners communicated with each other and with headquarters on the content of the services delivered, they were astonished at the differences. Other organizations have had similar experiences.

Two of the five consultants employed by this same company fall in the culture change school; the other three are in the worker-focused, behavior-based school. It is thus not surprising that dissimilarities between the services delivered are remarkable.

Unfortunately, the popularity of behavioral safety has enticed a large number of consultants to enter the field. Some of the more prominent consultants have, by implication, suggested that safety practitioners take care in assessing the credentials of the consultants they engage.

Comments made at the ASSE 1999 Professional Development Conference by Aubrey C. Daniels, a well-known and long-term practitioner in the field of performance management, are symbolic of developing apprehensions. The following is an excerpt from his paper, "The Imperative for an Integrated Approach to Behavior-Based Safety Initiatives":

> I have some rather serious concerns that many of the products and services being marketed under the name of behavior-based safety are not truly behavior-based, as in behavior analysis. My concern is that a lot of what I see going on in the name of behavior-based safety is not true behavior analysis, but what I call a watered-down variety of it (287).

Before a safety practitioner engages a consultant in behavioral safety, these three questions should be resolved: Is the approach taken principally intended to achieve an organizational culture change, or will the consultancy being offered principally focus on worker behavior, and what are the consultant's qualifications?

How does the culture change model differ from the worker-focused behavior modification model?

The Culture Change Model

How would one describe an organization's culture and what impact does that culture have on the hazards-related incident experience attained? In the author's *On the Practice of Safety*, this is the description given for an organization's culture:

> An organization's culture consists of its values, beliefs, legends, rituals, mission, goals, and performance measures, and its sense of responsibility to its employees, its customers, and its community, *all of which is translated into a system of expected behavior*.
>
> Management obtains, as a derivation of its culture, as an extension of its system of expected behavior, the hazards-related incident experience that it establishes as acceptable. For personnel in the organization, "acceptable" is their interpretation of what management does (18).

A much briefer description of an organization's culture is that it consists of the values for which management rewards and punishes. Consultants in the culture change school intend to help an organization change its safety culture, change its *system of expected behavior*, and help safety be considered a value for which rewards and punishments apply, among other important management elements.

Steven I. Simon and Terry E. McSween are among those prominent in the culture change school. Simon expressed this view at the 1998 ASSE Seminar:

> The culture drives the behavior and a culture change is required for success in behavior-based safety; whereas behavior modification is an individual-based training method in behavioral observation, analysis and feedback, culture change is a system-wide change effort (45).

McSween's comments at the same seminar focus on the significance of safety as a core value:

> We are striving to produce a culture where safety is a core value; if safety is a core value, the members of the culture will work safely all of the time, whether anyone is there to observe them or not; we need to understand the reasons people work safely or take risks and we need to create a work environment that supports people in working safely (44).

If a consultant is engaged who belongs to the culture change school, it should be expected that an early action for improvement will be to use a survey instrument, a form of perception survey, to assess the reality of an organization's culture with respect to safety: i.e., its system of expected behavior. This survey will include all levels of employment, with an emphasis on what senior management actually does. Its purpose is to present management with an analysis of the safety culture in place and recommendations for improving it. Only senior management can markedly change an organization's culture.

In the recently published *Safety Culture and Effective Safety Management*, Steven Simon has written a chapter entitled, "On the Future of the Safety Profession." He emphasizes that a culture which supports safety is a necessity for success:

> It is imperative for the safety professional to understand how culture affects organizations and to understand its impact on safety performance. Without a supportive culture, even the best designed safety processes or programs will fail (6).

The Worker-Focused, Behavior-Based Model

Aubrey Daniels, in the article cited earlier, credits Dan Petersen with introducing the term, *behavior-based safety*, to the profession in the late 1970s (288). And, it is Dan Petersen's definition of behavior-based safety, as in the proceedings of the ASSE 1998 Seminar, that will be used here:

> Perhaps the definitions we hear most about today are those that define behavior-based safety as a process of involving workers in defining the ways they are most likely to get hurt, thus getting their involvement and thus some buy-in, asking the workers to observe other workers to determine progress in the reduction of unsafe behaviors, etc. (2)

Obviously, worker-focused, behavior-based safety puts the spotlight on the worker and worker behavior. References in the literature to the work of B. F. Skinner, "the father of applied behavior analysis," are frequent on how to go about improving worker behavior.

This brief but helpful synopsis of Skinner's theory appears in Stephen H. Reynolds's "Back to the Future: The Importance of Learning the ABCs of Behavioral Safety":

> B. F. Skinner theorized that all behaviors are a function of "antecedents" and, perhaps to a larger extent, the "consequences" of those behaviors. Antecedents (also called "activators") serve as triggers to specific observable behaviors. Consequences either reinforce or discourage repetition of those behaviors. Most of today's behavioral safety movements are founded on this "ABC" theory (23).

A condensation cannot be given here of the methodologies presented in the many books available on applying the worker-focused, behavior-based safety methods that seek to achieve improvements by focusing on antecedents, behaviors, and consequences. As an introduction to those methods, consider this extract from *The Behavior-Based Safety Process*, 2nd edition, by Thomas R. Krause:

> Facilities that achieve continuous improvement as a safety goal do so by following certain steps and by treating each step as an opportunity to involve wage-roll employees (93).

These steps are then displayed in an *implementation sequence* that shows the order of the approach to be taken:

- Implementation planning meeting
- Assessment visit and report
- Behavioral inventory tools development
- Management training
- Ownership meetings
- Observer course development
- Observer course review
- Computer software training
- Observer training
- Kickoff meetings
- Ongoing observation and data collection
- Progress check
- Safety improvement process training
- Action planning, user conferences, benchmarking (95)

Krause's approach is quite a process. All of it is directed toward identifying and minimizing at-risk behaviors. Data would be gathered on at-risk behaviors through interviews with employees, safety surveys, and analyses of accident reports. At-risk behavioral inventories would then be produced. Worker groups would be trained to give each other positive reinforcement for safe behavior, the intent being that peer involvement might reduce at-risk behavior. Behaviorists have written that immediate, frequent, and positive rewarding of proper behaviors will promote the repetition of those desired behaviors.

Note again that the center of attention is the behavior of the worker. Rarely in the worker-focused, behavior-based literature is there recognition that the antecedents for at-risk behavior should be analyzed to determine their sources. Seldom is there recognition that the antecedents may derive from the work environment created by management: e.g., from the design of the workplace or the actual work methods.

Reynolds discusses the need to "first analyze the management controlled antecedents and consequences that actually drive those employee behaviors" (24). Many of the concerns expressed by highly capable safety professionals over the worker-focused, behavior-based safety approach center on the fact that causal factors deriving from the work environment and work practices are ignored.

For as long as some practitioners in behavioral safety continue to make speeches and write papers claiming that what they call behavioral science methodologies will solve all safety problems, the concerns voiced by knowledgeable safety professionals will continue to be substantive.

SIGNIFICANCE OF ROOT CAUSAL FACTOR ANALYSIS

There is a way to dispel the concerns of safety practitioners. Referring again to Krause, we cite his "steps of data-driven workgroup problem-solving":

1. Identifying a problem
2. Identifying root causes
3. Generating potential actions
4. Evaluating possible actions
5. Developing an action plan
6. Follow-up; measure and evaluate (79)

This list is, in effect, a good problem identification and resolution technique. (Kepner and Tregoe would be pleased with it.) And, if the practitioners in the worker-focused behavioral school applied such a problem identification and resolution process and documented this, they would largely dispel the concerns of safety practitioners who are skeptical about their methods.

Identifying root causal factors is an important safety process. In applying the process to identified at-risk behaviors, logic requires that a strong emphasis be given to distinguishing causal factors that are systemic from those that are truly behavioral. If the antecedents derive from the work systems that are the responsibility of management, then the first course of action should be to direct management attention to them.

However, the reality is that the need to conduct analyses to identify root causal factors has infrequently appeared in articles by practitioners of worker-focused behavioral safety. Further, it is common for advocates of that model to state that from 80 to 95% of all accidents are caused by unsafe acts, and, therefore, the proper action is to develop a worker-focused behavioral solution.

Although perhaps unintended, such a viewpoint creates the impression in the minds of managements and many safety practitioners that workers are the problem, that all risk situations can be resolved by worker observation techniques and positive reinforcement, and that the antecedents which derive from the design of the work methods and workplace can be ignored.

ANTECEDENTS DERIVING FROM THE PRESCRIBED WORK METHODS AND THE WORKPLACE

If a safety problem, in fact, lies within the system, attempts to reduce its risk through worker-related techniques will be only partially successful. Early on

The Behavior-Based Safety Process, 2nd edition, Krause makes the point that certain situations require engineering solutions:

> Yes, site personnel are to be congratulated on having discovered that standing in the line of fire is a critical behavior. Yes, they are to be supported in their effort to systematically measure and give feedback to their coworkers about the extent to which they expose themselves to risk by standing in the line of fire. However, in a case where the work configuration encourages or even requires the employee to stand in the line of fire, doing observation and feedback is necessary but not sufficient. Such a case is exactly where an engineering solution is called for, one that is directed by rigorously collected and well analyzed behavioral data. Stopping short of such system improvements can set the stage for workforce discouragement with the safety effort. Site personnel may well begin to doubt management's commitment to making genuine improvements to workplace safety if system improvement is lacking (5).

In *Techniques of Safety Management*, Dan Petersen offers a premise from which one must conclude that the antecedents for some at-risk behaviors are to be resolved by management:

> In most cases, unsafe behavior is normal human behavior; it is the result of normal people reacting to their environment. Management's job is to change the environment that leads to the unsafe behavior (35).

Management is responsible for the work environment created, the design of the workplace, and the design of work methods. If the antecedents for at-risk behavior derive, for example, from material-handling ergonomics problems that overstress a large percentage of the employee population, the focus for reducing at-risk behavior logically has to be on revising the material-handling methods.

Attempts to resolve identified ergonomics problems through behavior modification and positive reinforcement will achieve but limited results.

ONE LABOR UNION VIEW

Jim Howe is the Assistant Director of the Health and Safety Department at the United Auto Workers (UAW) Union head-quarters. He is often a speaker on behavioral safety and takes a strong position, insisting that "behavior-based safety can be hazardous to your health and safety program." Why so? His views are presented in "A Union Critique of Behavior-Based Safety," a paper issued by the UAW Health and Safety Department (10). The author believes that the following are the key points in the union critique and Jim Howe's position, and that thoughtful safety practitioners should give them careful consideration:

- Behavior-based safety is based on one of the oldest and most outdated approaches to safety—the Heinrich premise that 88% of all industrial accidents are caused primarily by unsafe acts of persons (1)....

- Behavior-based safety does not focus on what really causes accidents.

  Injuries and illnesses are caused by exposures to hazards. What do I mean by hazards? Hazards include any aspect of technology or activity that produces risk. If the work methods designed and prescribed put employees at risk, those work methods are hazardous" (5)....

- Behavior-based safety does not recognize the most effective method of controlling hazards, which is through the application of the hierarchy of controls.

  The hierarchy recognizes that design, elimination and engineering controls are more effective in reducing risk than lower level controls such as warnings, training, procedures and personal protective equipment.... (Behavior-based safety practitioners) do not support it because it contradicts their theory that 95% of accidents are caused by unsafe acts (7)....

- Behavior-based safety places responsibilities on workers for which they may not be qualified.

  Although worker involvement is important, it has limitations and is *not* a substitute for technically competent health and safety experts reviewing both existing and future operations to insure that hazards are identified and controlled. Few workers have been trained in hazard identification, risk evaluation or methods of control (hierarchy). Workers rely on their past history, which may be quite limited, for ideas about possible methods to control hazards" (17).

The same paper has the following to say about the hierarchy of health and safety controls:

> This hierarchy recognizes that design, elimination, and engineering controls are more effective in reducing risk than lower levels such as warnings, training, procedures, and personal protective equipment. In a descending order of effectiveness, the hierarchy given is this:
>
> 1) Elimination of Substitution
> 2) Engineering Controls
> 3) Warnings
> 4) Training and Procedures, Administrative Controls
> 5) Personal Protective Equipment (7)

Reference to a hierarchy of controls are made in two places in the remainder of this chapter. Those references are clear indications of the transitions in progress in the field of behavioral safety.

In the author's discussions with Howe, it became apparent that he has read extensively on behavioral safety. For as long as the behavior-based safety literature ignores injury causal factors that derive from (as cited above) "work methods designed and prescribed (that) put employees at risk," it can be expected that Howe and other union representatives will continue to make the following case: Behavior-based safety is not what some of its practitioners make it out to be. One of Howe's concerns is the impression being made by many regular presenters at professional meetings, who continue to suggest that behavior-based safety is all that practitioners need to focus on.

Some of these proponents have given talks in which they claim that 85 or 90% of all accidents are caused principally by unsafe acts. No matter what definitions are given for the term unsafe act, they immediately move to addressing means for resolving the at-risk behavior of employees through behavior modification methods, with minimum or no consideration of systems causal factors. To the author, this approach is absurd. Those who continue to insist that a vast majority of accidents are caused principally by unsafe acts do the work world a disservice.

THE PROPER ROLE FOR BEHAVIORAL SAFETY

Michael E. Porter is the Global Manager of Safety, Health & Worker's Compensation at The Goodyear Tire & Rubber Company. He was a presenter at a National Safety Congress session in October 1999. His presentation described how to work with union representatives when establishing a behavior-based system, and more important, where behavior-based safety fits within an overall safety initiative.

Behavior-based safety was introduced at Goodyear to complement existing systems. To encourage union support, Goodyear communicated early in the process with union representatives on the role of behavior-based safety within the company's safety plan. These are excerpts from Goodyear's letter to Jim Frederick at United Steel Workers of America (USWA):

- Behavior-based safety is only "one" part of our "safety balance" model. Behavior-based safety was selected by Goodyear to further enhance and support existing safety and health systems.
- Systems, conditions, and behaviors are equally important to injury and illness prevention.
- Ergonomics, engineering, safety committee activity and other safety systems remain active.
- Behavior observation is not used for discipline. Observation data is used by problem solving teams to identify and correct systems, procedures, and design issues which encourage at-risk behaviors.
- Hourly and salaried buy-in is required prior to implementation.

What is the message here? Behavior-based safety was included as a complement to existing safety systems to improve safety results. Effective safety systems were continued. Contrary to what some speakers suggest, it was not assumed that behavior-based safety, by itself, would solve all safety issues. A strong focus was maintained on factors such as systems improvement, ergonomics, engineering, and safety committees.

Of great significance was the outcome of the identification of at-risk behaviors. In many of the situations described by Porter during his presentation, analysis of the at-risk behavior to determine their antecedents resulted in real problem solving, i.e., in the redesign of a workplace or work methods. The following are some of the modifications mentioned by Porter: installed bump bars, eliminated pinch point, ergonomic situations corrected, barriers installed, feed and exit conveyor redesigned, potential crush injury eliminated.

If behavior-based safety is to complement a well-established safety initiative and the identified at-risk behaviors are analyzed to determine the nature of their antecedents, then it can well serve to reduce the risk of injury.

TRANSITIONS IN BEHAVIORAL SAFETY AS EXPRESSED BY ONE PROMINENT LEADER

Thomas R. Krause, Chief Executive Officer of Behavioral Science Technology, was also a presenter at the National Safety Council session at which Porter was a speaker. He addressed the management and labor perspectives of behavior-based safety. A few of his observations are cited here:

- A good attempt should be made to prompt union leadership buy-in before implementing behavior-based safety.
- Both management and union personnel should understand that behavior-based safety is not a substitute for good engineering, facilities, training, and the use of the hierarchy of controls.
- Emphasis should be given to the effect behavior-based safety has on accomplishing goals, improving safety systems, and problem solving.

To reiterate: Krause said that behavior-based safety is no substitute for good engineering, or the use of the hierarchy of controls. These are significant comments, coming from a prominent leader in behavioral safety. Safety practitioners should give careful consideration to what these comments portend.

Krause provided extensive statistics reflecting the research done by Behavioral Science Technology to establish the success of well-implemented behavior-based initiatives. Some of them are presented in Table 1.

TABLE 1 Percent of Identified Barriers to Safe Behavior

Subject	Percentage (%)
Management system	13
Rewards/recognition	1
Facility and equipment	60
Disagreement on safe practices	2
Personal factors	2
Culture	8
Low adaptive readiness	7

Speech by Thomas R. Krause, CEO, Behavioral Science Technology, National Safety Congress, October, 1999

Barriers to safe behavior deriving from the culture, management systems, and facilities and equipment total 81%. That suggests the greatest risk reduction will result from attention to those three subjects, and that an early emphasis following the identification of at-risk behaviors should be to determine the reality of their causal factors, the antecedents. The author finds such an approach most interesting. Does this suggest the next step in the evolution of the applied practice of behavioral safety?

All this ties in with the actions taken at Goodyear to reduce at-risk behavior whereby modifications were made to systems, the design of the workplace, and actual work methods.

Another part of Krause's presentation was also of major significance. W. Edwards Deming's reputation in quality management spans the world. The use by Krause of quotes from Deming's work suggested that he had reoriented his approach to behavior-based safety in relationship to some of Deming's concepts. As an example, Krause quoted the following:

> Performance does not come from the individual ... but from the system.

This recognition by Krause of a significant premise that permeates Deming's writings is important. Since performance derives from the system, and not the individual, the proper outcome of behavior-based safety should be to improve the system. Accepting that premise would be a bit revolutionary for the worker-focused, behavior-based safety consultants who have stressed behavior modification and positive reinforcement as the solution to safety problems.

Deming's premise relates directly to Table 1. Again, we find the connection very interesting, for Krause presented a markedly different role for behavioral safety in his overall talk.

In a later presentation at the ASSE's 2000 Professional Development Seminar, Krause carried his ideas one step further:

- The behavioral process is a method for continuously improving facilities, equipment, design, and management system issues.
- Behavioral-based safety should not be regarded as the entire safety system, but rather as one component of many.
- The situation in which the injury is actually *caused by the worker*, while existent, is extremely rare.
- Stating that 80 to 95% of accidents are caused by unsafe acts is wrong, because the primary causes result from management systems and the facility.
- The great majority of the actual causes of the injury is an interaction between the worker and the facility.
- This point of interaction is the working interface, which should become the subject of attention.
- Determining how the worker interacts with the system changes the focus of improvement from the worker to the systems that enable safe behavior.

To say that the focus of improvement is not the worker, but the systems that enable safe behavior is a major conceptual change in behavioral safety. Can it be doubted that Krause expressed a substantial shift in behavioral safety concepts?

ANOTHER INDICATION OF TRANSITION

Earl Blair's article entitled, "Behavior-Based Safety: Myths, Magic, Reality," is additional recommended reading. When that article was written, Dr. Blair was associated with Safety Performance Solutions, of which Dr. E. Scott Geller is a senior partner and founding member. Blair lists 12 "myths" about behavioral safety and his arguments against them. Consider the following excerpts:

- **Myth 1: Behavior-Based Safety Doesn't Address System Causes of Injuries**
  Reality: Behavior-based safety can and should address system causes. In fact, it is an effective way to determine root causes of injuries, at-risk behavior or hazardous conditions. Once at-risk behaviors/conditions are identified, system barriers to safe work performance can be identified and removed (26). . . .
- **Myth 4: Behavior-Based Safety Has Become Myopic**
  Reality: Some critics contend that behavioral-safety has become myopic. Is this criticism True? It can be if applications are narrow and misguided. Behavior is just one aspect of safety. In the process of focusing on behavior, other key components—such as engineering, monitoring and machine guarding—should not be ignored. .Efforts to reduce injuries must be comprehensive (27). . . .
- **Myth 6: Behavior Interventions Are the Least-Effective Interventions**
  Reality: Most safety professionals agree with the traditional safety hierarchy: 1) engineer the hazard out; 2) control the hazard; 3) guard against the hazard; and 4) use training and personal protective equipment. Even though the behavioral component is low in this hierarchy, it must not be overlooked (27).

Such a point of view is fascinating and so different. For a moment, think about what Blair wrote: Behavior-based safety should address system causes; root causes should be determined; behavior is one aspect of safety; and there is agreement with the traditional hierarchy of controls. Does all of this indicate the repackaging of a product?

CONCLUSION

Events to come will dispel the confusion about what behavioral safety is and who really practices it. Quite certainly, there are transitions in progress that place behavioral safety properly in the context of an overall safety initiative.

In *On the Practice of Safety*, the author states the following:

> This essay pleads for a balanced approach that gives a proper emphasis within the practice of safety to causal factors deriving from design management, operations management, and task performance (67).

Without question, there is a place within the practice of safety for both the culture-driven behavioral safety model, which is to improve a safety culture and achieve a well-structured and managed safety initiative, and a worker-focused, behavior-based approach that augments a well-established safety system.

In either case, consideration should be directed to the real sources of risk problems, whether they be culture- and management-system-based, or worker-behavior-based, or both. If the risk problem lies within the system, a culture and systems modification solution should be sought. If task performance problems are truly behavioral, consideration should be given to resolving those problems through prudently selected behavior modification techniques. For most risk situations, the causal factors and methods of resolving them will be multifaceted.

This author wants the best for both employers and employees: A culture-driven safety initiative in which the expected behavior, established by senior management, requires superior policies, standards, and procedures impacting optimally on design management, operations management, and task performance. And, an effective accountability system for their application.

REFERENCES

Blair, Earl. "Behavior-Based Safety: Myths, Magic, Reality." *Professional Safety*, Aug. 1999.

Daniels, Aubrey C. "The Imperative for an Integrated Approach to Behavior-Based Safety Initiatives." In *ASSE 1999 PDC Proceedings*, Des Plaines, IL.

Geller, E. Scott. "Confusion, Controversy, and Clarification." *Occupational Safety and Health*, Jan. 1999.

Krause, Thomas R. *The Behavior-Based Safety Process*, 2nd ed. New York: John Wiley & Sons, 1997.

Manuele, Fred A. "Perspectives on ASSE's Behavioral Safety Symposium." *Professional Safety*, Aug. 1998.

Manuele, Fred A. *On the Practice of Safety*, 2nd ed. New York: John Wiley & Sons, 1997.

Petersen, Dan. *Techniques of Safety Management*. Goshen, NY: Aloray, Inc., 1989.

Proceedings: A Behavioral Safety Symposium. Des Plaines, IL: ASSE, 1998.

Reynolds, Stephen H. "Back to the Future: The Importance of Learning the ABCs of Behavioral Safety." *Professional Safety*, Feb. 1998.

Simon, Steven. "On the Future of the Safety Profession." In *Safety Culture and Effective Safety Management* (George Swartz, Ed.). Itasca, IL: National Safety Council, 2000.

"A Union Critique of Behavior-Based Safety." Version 2, UAW Health and Safety Dept., Detroit, MI, 1998.

11

The Safety Through Design Movement

INTRODUCTION

Developments reviewed in Chapter 5, "Why You Need to Know About Hazard Analysis and Risk Assessment," are indicative of an emerging safety through design movement. There is an irony to this fairly recent development since safety literature dating back over 70 years has clearly established that the most effective way to reduce risks and avoid injuries and damage is to have hazards recognized and risks properly addressed during the design process. This chapter will assist safety practitioners in understanding what to expect concerning the adoption of safety through design concepts and how those concepts will affect the substance of their work.

THE INFLUENCE OF EUROPEAN STANDARDS

European safety and quality standards issued in the past 15 years have had a worldwide influence on the design of machinery, equipment, and consumer products. How broad is that influence? Countless articles and books have been published pertaining to those standards. This excerpt is from one of the many articles that have appeared in recent years indicating how great an impact the Europeans have had. As Peter M. Engstrom wrote in "Designing Machinery to Meet Global Safety Requirements":

> The global harmonization of safety standards, and the desire of machinery manufacturers to produce "world" products, has resulted in the adoption of many of

> the machine guarding concepts embodied in the European Machinery Directive (EMD).
>
> Both domestic and international standards-making organizations that influence machine design are using the European Machinery Directive as a point-of-reference in developing new guidelines (1).

Safety practitioners should expect to see over time the worldwide influence of European safety requirements on standards-making organizations. European safety standards require that hazards analyses and risk assessments be made during the design process. Making hazards analyses and risk assessments early in the design process is one of the premises on which safety through design concepts are built. Including such provisions in American standards and guides is already evident, as discussed in the publications that are the subject of Chapter 5.

And, Europeans are moving forward to achieve a more widespread application of hazard analysis and risk assessment methods. They are actively promoting the development of a generic standard that addresses these subjects. Such was the focus of a conference held in Italy in May 2000 at which international attendance was notable.

GLOBAL HARMONIZATION OF SAFETY AND QUALITY STANDARDS

The beginning of the global harmonization of safety and quality standards is a fact. Prominent evidence of it may be observed in the pervasive adoption throughout the world of the quality management and assurance standard known as the ISO 9000 series issued by the International Organization for Standardization (ISO). The ISO is the specialized international agency for standardization. At the time of this writing, the national standards organizations of 91 countries participated in the ISO. The American National Standards Institute (ANSI) is the member body representing the United States.

The ISO 9000 series is a set of five individual, but related, international standards on quality management and quality assurance. They are generic, not specific to any particular products or industry. The title of the ISO 9000 series is "Quality management and quality assurance standard—Guideline for selection and use." The United States had input in the development of the ISO 9000 series through membership in ISO via ANSI and adopted the ISO 9000 series word for word as the ANSI/ASQ Q9000 series. (ASQ stands for the American Society for Quality, which was the American Society for Quality Control.)

Corporations throughout the world engaged in international business have built and continue to build their quality systems around the ISO 9000 series. It can be expected that more purchasers of products and components will require that suppliers meet or exceed the requirements of the standard for quality and safety.

For example, the Automotive Industry Action Group formed by DaimlerChrysler, Ford, and General Motors has published Quality System Requirements (QS-9000). They are a close adaptation of the ISO 9000 series. All suppliers of automobile parts and components to these companies are required to meet their quality system requirements, one of which is to make a failure modes and effects analysis (FEMA). Other industries have also published their own adaptations of the ISO 9000 series.

Worldwide acceptance of the ISO 14,000 series of standards on environmental management, adopted in 1996, is another indication of the safety through design efforts emanating from Europe. This is what the National Safety Council said about these standards in its *Accident Prevention Manual for Business and Industry*:

> They will dramatically change and broaden the ways in which environmental management is conducted and, potentially, regulated in the global economy (300)....
>
> The safety/health/environmental professional must be familiar with the most current environmental regulations. The ISO 14,000 and similar regulations are providing international business with stronger guidelines and regulations for protecting the environment and safeguarding worker and consumer health and safety (309).

European standards pertaining to the safety of machinery that goes into the workplace and the risk assessments they require are discussed in Chapter 5. Throughout the world, manufacturers of machinery and equipment to be sold to European employers must meet the requirements of the ISO standards.

Furthermore, member countries of ISO have promoted the drafting of an international safety management standard, for which discussion and meetings have already taken place. Such a standard will be issued, in time.

European standards will continue to have a global influence and safety practitioners can expect to see the further incorporation of their hazard analysis and risk assessment requirements in ANSI standards and industry guidelines. They require addressing hazards and risks early during the design process.

AN INTERESTING UPTAKE ON SAFETY THROUGH DESIGN

The first edition of the *Standard Handbook of Machine Design* was issued in 1986. When consideration was given to a second edition, which was published in 1996, it was determined that two new chapters were needed: a chapter on usability and another on safety.

The chapter on usability is, in fact, a dissertation on incorporating ergonomic considerations in the design process. Sections of the chapter have headings like "Designing for Human Body Size," "Designing for Human Body Posture," and "Designing for Reach and Mobility."

In the chapter on Safety, a general review is given of why safety, what is safety, hazards and risks—which is followed by a review of the designer's obligation for integrating safety considerations into the design process.

Shigley and Mischke, editors of the second edition, had clearly recognized the necessity of giving greater importance to designing for safety. Also, in a revised first chapter, they in a sense spoke of the emergence of the safety through design movement: "The inclusion of safety and reliability at or near the level of importance of function is a recent development (1.2)."

OTHER APPLICATIONS OF SAFETY THROUGH DESIGN PRINCIPLES

Intel Corporation issued an *Environmental, Health and Safety Performance Report* in April 2000. Excerpts from it indicate just how extensively this organization has embraced safety through design principles:

> **Intel's EHS Organization**
>
> An overarching strategy and one of the keys to our success has been the integration and proliferation of EHS improvements in manufacturing developments efforts. This is a strategy we refer to as Design for Environment, Health and Safety (DFEHS). The success of DFEHS processes relies on management and manufacturing commitment, and a broad understanding of EHS impacts, from tool and supplier selection to facility design and waste and resource management.
>
> EHS staffs lend direction as members of manufacturing and product teams chartered with technology development, process design, product ecology, chemical and natural resource strategy, and facility design (8). . . .
>
> **Design for Environmental, Health and Safety**
>
> The integration and proliferation of EHS improvements in the way we manufacture our products require a comprehensive understanding of tool and supplier selection, facility design, waste and resource management—and the way they effect EHS results.
>
> Our very detailed analysis of such factors allows us to set EHS performance criteria for five generations of future process technology, or 7 to 10 years ahead. It also allows us to build EHS improvements into the design process, thereby improving our results and avoiding expensive EHS-related retrofits (10). . . .
>
> **Safety Performance—Equipment Design for EHS**
>
> Achieving cost-effective compliance with EHS requirements has become increasingly difficult with the rapid pace of modern technology cycles and the continued globalization of Intel's operations. Nevertheless, Intel has met this challenge by asking its key equipment suppliers to integrate a DFEHS model into their existing business methodology. As within Intel, this supplier model incorporates EHS requirements and analyses into the tool development, starting at the design concept and continuing through each subsequent phase. Competitive advantages include lowered cost of compliance and further reductions in injuries and other incidents among workers (13).

Consider this excerpt from a communication to the Project Manager of the Institute for Safety Through Design on what one entity has accomplished in terms of safety through design:

> The Operating Engineers National Hazmat Program has been working hard to see that safety is integrated into the design of new remediation technologies developed by the Department of Energy. We have conducted safety and health evaluations of over 40 technologies and have more than enough industrial hygiene data and safety observations to draw some clear lessons. We will be hosting a technical workshop that will focus on making a clearer case for the economic benefits of integrating safety into the design or remediation technologies.

This last example of the application of safety through design principles is an ad for employment. In the June 2000 issue of *Professional Safety*, the magazine of the American Society of Safety Engineers, Kodak Company advertised for a safety engineer who would have "experience in applying systems, safety and ergonomic design principles in the early design of manufacturing processes and facilities ... and lead a safety through design program."

These examples are cited as the possible basis for reflection by safety practitioners on what will be required in their work in the future, and of the opportunities this transition presents.

LEADERSHIP FROM THE CHEMICAL INDUSTRY

Bruce Main, President of Design Safety Engineering, partially fulfilled his graduate degree by conducting a study of and presenting a thesis on the following topic: "What Do Design Engineers Really Know About Safety?" Although his study pertained to the mechanical engineering design community, it has been found that one of his conclusions is typical of most engineering disciplines. Main observed that, "Design engineers receive little or no formal safety education and training (iv)."

Members of the American Institute of Chemical Engineers made a similar observation. An Undergraduate Education Committee of its Center for Chemical Process Safety formed SACHE, which stands for Safety and Chemical Engineering Education. Its purpose is this: Prepare and distribute educational materials to university members to assist them in adding safety, health, and loss prevention to the chemical engineering curriculum.

Much of its work, represented by instruction modules, videos, and presentation text, relates to designing for safety. Among the engineering societies, chemical engineers stand in a leadership position in terms of promoting safety through design concepts.

Now, just what is safety through design and why should its concepts be applied?

DEFINING SAFETY THROUGH DESIGN

This is the definition of safety through design that was agreed upon by the Advisory Committee to the Institute for Safety Through Design:

> Safety through design is defined as the integration of hazard analysis and risk assessment methods early in the design and engineering stages and the taking of the actions necessary so that the risks of injury and damage are at an acceptable level.

Take note that this concept encompasses facilities, hardware, equipment, tooling, materials, layout and configuration, energy controls, environmental concerns, processes, and products. Whatever the consideration—safety, health, or the environment—the applicable principles and methods are the same.

What is proposed for safety through design is an agreed upon and well-understood concept: a way of thinking that is translated into a system which effectively addresses hazards and risks in the design process.

THE SAFETY THROUGH DESIGN MODEL

Figure 1 presents the safety through design model. In applying the concept, emphasis is given to moving consideration of hazards and risks "upstream" to the beginning of the design process, rather than what is often the typical sce-

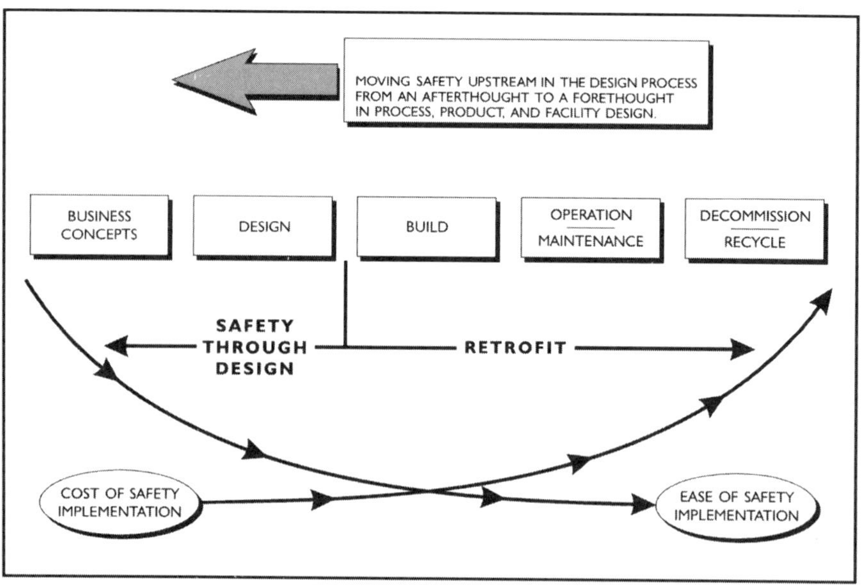

FIGURE 1 *Moving safety upstream in the design process.*

nario, whereby safety considerations are really an afterthought. Upstream includes all aspects of the original business concepts, relative decision making, and design process.

Integrating a consideration of hazards and risks early on in concept exploration, business decision making, research, and design results in easier, more effective, and less expensive safety implementation. Also, costly retrofitting during the build, operation, maintenance, and decommissioning periods is avoided. The model demonstrates that the cost of implementing safety increases as a project moves beyond the design stage.

Although the same hazard analysis and risk assessment methods may be used while effecting risk reduction modifications in the build, operation, maintenance, and decommissioning phases, those modifications are, in reality, retrofitting.

BENEFITS TO BE OBTAINED

If safety through design concepts are applied, what is to be gained? The benefits of integrating decisions affecting safety, health, and the environment in the early stages of the design will include:

- Significant reductions in injuries, illnesses, and damage to the environment, and their attendant costs
- Productivity improvement
- Decreased operating costs
- Avoidance of design shortcomings and expensive retrofitting

Anecdotal data are abundant indicating that doing things right to begin with results in reduced injuries, illnesses, and damage to the environment; improved productivity; and lower operating costs. Ergonomics provides the example to be given here of the recognition that substantive gains will be achieved when hazards are eliminated during the design process.

A dissertation on relatively new innovations to eliminate ergonomic risks in the design process appears in an article entitled, "Chrysler's Ergonomics Program Enters the Fast Lane." One pull-quote reads, "Chrysler Corporation's operating system places ergonomics in the front seat of the design-build process. It means there will be no ergonomics surprises at launch time."

This article states that working with CATIA software, engineers are now able to include ergonomics in the earliest stages of vehicle development. (CATIA stands for computer-aided, three-dimensional interactive application, a software program created by Dassault Systemes.) Furthermore, the article states that ergonomics problems are being considered in tool designs and building processes, along with other aspects of the development process. Simulation in the software program, the article attests, helps find a solution to any problem created in the design process.

According to the article, reducing jobs that create cumulative trauma disorders is a high priority. It further says:

> Quantitative data now available prove that solving ergonomic problems improves product quality as well as productivity of the work force (43).

This article is but one indication of the progress being made in developing software that assists engineers in addressing safety needs as systems and products are being designed.

SOFTWARE FOR SAFETY THROUGH DESIGN

The software program used by Daimler-Chrysler is proprietary, and some others say they have their own proprietary programs. Thus far, no software developer has publicly offered a program that incorporates safety through design concepts. However, the interest in developing such software programs is growing.

In an article entitled, "The Challenge: Safety through Design," Wayne Christensen, Project Manager for the Institute for Safety Through Design, reviewed the safety through design process, working through the model. He presented a challenge to the software industry to proceed with the development of publicly available computer-assisted design software that would accelerate moving safety upstream in the design process.

CATIA placed the article on its website, where it has been frequently visited. Although CATIA speaks to the needs of one set of design engineers, two other magazines, addressing different design engineering skills, have also run the article. Developments are occurring, of which safety practitioners should take note and an interest.

At a recent presentation to the ISTD Advisory Committee, a CATIA applications engineer with Dassault Systemes demonstrated the capability of "CCPlant" software to consider safety, health, and environmental concerns in the design process. To add to their network of software solutions, Dassault Systemes recently purchased SAFEWORK, a software program that has the capability to assess ergonomic risks to workers who are to manufacture or assemble a product while it is being designed.

ATTENTION TO AVOIDING RETROFITTING COSTS

In the discussions resulting in the development of the safety through design model, much was made by some participants of the distinction between designing things right in the first place and retrofitting to make corrections after designs had been agreed upon. Companies represented by those participants had found that retrofitting costs—whether for a facility, process, or product—could be enormous.

Richard Parker of Intel gave an example of how costly retrofitting can be in a 1999 presentation at the National Safety Congress. Parker said that the cost of retrofitting at Intel for ergonomics in 1 year was $32,000,000. That cost was reduced over 3 years to less than $100,000 annually by insisting that suppliers of equipment and machinery adhere to clearly established ergonomic specifications in their design processes.

Safety practitioners should appreciate this opportunity for recognition if the counsel they give to management on how to avoid bringing hazards into the workplace results in avoiding costly retrofitting.

Mention of the Institute for Safety Through Design has been made several times here. Safety practitioners should be aware of what this organization is and what it has accomplished.

THE INSTITUTE FOR SAFETY THROUGH DESIGN

The National Safety Council created the Institute for Safety Through Design (ISTD) in 1995. Its origin arose out of the recognition by safety professionals that great strides forward could be made with respect to safety, health, and the environment if proper consideration was given to hazards and risks during the design process. A high-level group of individuals dedicated to the concept of safety through design agreed to participate in an ISTD Advisory Committee to provide overall direction. This is the mission of the institute as adopted by its Advisory Committee:

> To reduce the risk of injury, illness and environmental damage by integrating decisions affecting safety, health, and the environment in all stages of the design process.

In the institute's literature, it is made clear that, for simplicity, the word safety includes the broad areas of occupational safety and health, environmental concerns, and the safety of products.

> The strategy adopted by the Institute early in its development, and which has guided its activities indicate that it will:
>
> - *Conduct symposia and seminars to expand the knowledge and concepts of safety through design.* The Institute held a symposium in 1996 to define the scope of safety, health, and environmental knowledge an engineer should have upon successful completion of a baccalaureate degree program. Discussions of educators, engineering managers, and safety professionals during this symposium were valuable and varied. A COMPENDIUM is available that includes their presentations.
>
>   At a 1997 symposium on "Benchmarks for World Class Safety Through Design," papers were presented from general industry, as well as electronics, automotive, aircraft, health care, chemical, and construction industries. PROCEEDINGS for that symposium were published and are available.

Members of the Advisory Committee have participated in numerous conferences as presenters to promote safety through design concepts. For example: the Institute has conducted several one-day, professional development conferences at the annually held National Safety Congress; and the Institute has been the sponsor of technical presentations during the Congress.

- *Develop course content, study problems, case studies and other materials to enhance engineering curriculum on safety through design.* A program has been undertaken in a joint venture with the State of Utah and the University of Utah to develop undergraduate safety and ergonomics materials to be integrated into the mechanical engineering curriculum. It is not proposed that a new and separate course in safety be developed. Rather, the materials will be embedded in existing courses as additional problem sets, homework, and lab assignments.

  To influence the content of senior capstone design courses in engineering curricula, the Institute purchased copies of the software tool "designsafe ... the hazard analysis and risk assessment guide" and established a grant program through which the software is provided without cost to engineering departments that meet certain criteria for its use. As of this writing, ten engineering departments in nine engineering schools have met the criteria and have been awarded grants.

  A significant undertaking was the publication by NSC Press in July of 1999 of a book, appropriately titled *Safety Through Design*. The Project Manager for the ISTD, Wayne Christensen, and Fred Manuele were co-editors. Others advised us that we would have difficulty getting authors to write the chapters for which subject and content outlines had been prepared. The growing interest in safety through design proved those advisors false. Only one prospective author who was asked to write a chapter declined participation.

  The book provides a range of perspectives from twenty authors who share their knowledge and expertise in safety through design. There are three major sections in the book:

  1. Introduction to Safety Through Design;
  2. Integration Into Business Processes; and
  3. Safety Through Design in Industry

  Throughout the book, encouragement is given to individual company customization in moving safety upstream in the decision-making processes. Part three addresses safety through design in general industry and in several specific industries.

  Response to the book has been good: in less than six months, a third printing was necessary

- *Establish liaisons with engineering schools, professional societies, industry and labor to increase awareness of the Institute and its Mission.* The Institute has established a Coalition of universities to promote the inclusion of safety, health, and environmental knowledge in engineering undergraduate curricula.

The foregoing represents some of the work of the Institute for Safety Through Design. Additional information about the institute, or involvement with it, is available from Wayne Christensen, Project Manager, Institute for Safety Through Design, P.O. Box 303, Crystal Lake, IL 60039-0303. Or visit the NSC website at http://www.nsc.org/ISTD.htm.

APPLYING CONCEPTS IN THE REAL-WORLD PRACTICE OF SAFETY

Understandably, a safety practitioner may ask the following question: How do safety through design concepts apply to the work I do on a day-to-day basis? To answer that question, data follow on how the concepts are applied in avoiding specific injury types, followed by comments on broader applications of the concepts in certain industries.

Fall Prevention

In a very large company, 36 employee fatalities occurred as a result of falls over a 20-year period; two-thirds of the fatalities were skilled trade personnel. A team consisting of engineers, safety practitioners, and union safety representatives was formed to study the problem. As a result of their efforts, no fatalities from falls have occurred in the past 9 years. The study group applied the hierarchy of controls, understanding that the best alternative was to eliminate the hazards through workplace and work methods design or redesign, and that engineering controls would be the next best option.

What sort of actions were taken as a result of the study made? It became obvious early on that the potential for falls arose largely from the way things were designed, requiring a too frequent need to use ladders, particularly to do maintenance work. To minimize the use of ladders, shut-off valves and electrical disconnects were moved to ground level, and when the objects to be maintained could not be moved, work platforms accessed by fixed stairs were installed. When ladders were still necessary, fall protection devices were used when appropriate.

Although much of this work was retrofitting, a result of the study was the incorporation of specifications into the design requirements for new or altered facilities that provide ready access for maintenance people, and a minimum use of ladders.

Noise Elimination and Control

A good example of the benefits that can be achieved from applying safety through design concepts is described in the following excerpts from an article entitled, "Sound Advice for a Quieter Workplace." For this article, the subcaption is "Companies take a proactive stance against noise with engineering controls."

> Over months and years, the total costs of replacing equipment and the annual testing of every exposed employee can become significant, both in dollars and in time. Finally, when everyone is wearing hearing protection, *errors and delays in communication might cause costly losses in quality and productivity*. (Emphasis added.)
>
> The limited effectiveness of administrative controls and testing has led many companies to the conclusion that engineering controls offer the most reliable and lasting way to reduce workplace noise.
>
> The most effective time to apply engineering controls is before a noise source ever comes into the workplace. Establish noise standards and work closely with suppliers at the time machinery is specified and ordered. Ford's noise standard, SX-1, has been in effect for more than 20 years and is routinely included in the specifications for new and rebuilt machinery....
>
> (When a Ford plant was reconstructed) noise was eliminated by design.... Ford's efforts have paid off. Recent tests that equipped 300 employees in the plant with noise dosimeters showed none were being exposed to noise above 85 dBA. Of 700 employees in the new plant, fewer than 60 are required to wear hearing protection or undergo testing....
>
> ... Any company that buys equipment can specify maximum noise levels for new purchases (48).

This is a good depiction of what safety through design is all about: identifying the hazards and assessing the risks early in the design process, and designing and engineering to eliminate or control the risks. (The author first cited the previous excerpts in *On the Practice of Safety*.)

Avoiding Electrocutions: Lockout/Tagout

Edward V. Grund gives a thorough history of fatalities and injuries resulting from unexpected energy transfer in the second chapter of *Lockout/Tagout—The Process of Controlling Hazardous Energy*. One of the references Grund cites is the 1983 study made for NIOSH entitled, "Guidelines for Controlling Hazardous Energy During Maintenance and Servicing," which was one of the resources that gave impetus to OSHA issuing its standard on the control of hazardous energy sources.

Unfortunately, electrocutions continue to occur. During discussions about an electrocution in which the author participated, a safety director stated that just complying with the OSHA standard will not necessarily overcome electrocution risks if the lockout/tagout system in place induces workers to break the rules. In the investigation of the incident, the principal causal factor was found to be the fact that the electrical disconnect was an extraordinary distance away from the work.

In that company, additional disconnects are being installed, and design specifications have been rewritten to apply to new and altered operations. Unfortunately, the lockout/tagout system redesign and the rewriting of specs took place after the fatality occurred.

It should also be understood that merely requiring design specifications to state that an electrical system shall meet the requirements of the ANSI National Electrical Safety Code will not assure that disconnects are readily accessible.

Other safety directors have also indicated difficulty in getting workers to follow the lockout/tagout procedures prescribed. They too have recognized that, in some instances, the design of the systems in place is error-provocative. Safety practitioners, take heed: If the design of a lockout/tagout system does not make electrical disconnects readily accessible, an error-provocative situation is created for which it can be reasonably assured that errors will occur. Obviously, this is an initial design and redesign situation in which safety practitioners can become influential.

Preventing Musculoskeletal Injuries

Statistics indicating that, within the entire spectrum of injuries and illnesses resulting in absences from work, strains and sprains maintain prominence. A Bureau of Labor Statistics Report dated April 22, 1999, covering the 1997 reporting year, makes these statements:

- Four out of 10 injuries and illnesses resulting in time away from work in 1997 were sprains or strains, most often involving the back, as in the preceding 4 years. The number of sprains or strains cases declined nearly 17% from 1993 to 1997, representing slightly less than the decline for all cases (19%).
- Strain and sprain was, by far, the leading nature of injury and illness in every major industry division, ranging from 32% in agriculture, forestry, and fishing to 52% in transportation and public utilities.
- Overexertion while maneuvering objects led all other disabling events or exposures, cited in a sixth to a third of the cases in every industry division.

The literature is abundant suggesting that back injuries are costly. In a study of incident investigation reports made by the author, it became apparent that far too many back injuries are "programmed as nearly certain to occur" because the work methods prescribed overstress too large a percentage of the work force. For example, and this is a true story: A recent newspaper employment ad stated that the applicant had to be able to lift 120 pounds. That describes the type of situation in which back injuries are virtually "programmed."

If a safety practitioner wants to reduce the frequency and severity of back injuries, a worthwhile course of action is to study the work methods in the areas where the accidents have occurred to determine whether certain aspects of workplace and work methods design are, in reality, the causal factors. Many safety practitioners have done this, with remarkably good results. Often, they found that the fixes needed were not overly expensive.

Earlier in this chapter, mention was made of the computer software program at Daimler-Chrysler that addresses ergonomics in the design process. That innovation should be appreciated for the substantial benefit it provides in reducing the frequency of musculoskeletal injuries. Also, it can be presumed, logically, that in the successful avoidance of ergonomics retrofitting costs achieved at Intel by insisting that equipment suppliers meet prescribed design standards, a safer workplace resulted.

As one colleague observed, a safety practitioner should not assume that even though an equipment supplier may insist that ergonomics risks were considered in its design process that the standards applied were sufficiently high. He found that an equipment supplier thought it was acceptable to design for the capabilities and limitations of 50% of the employment population. It took revised specifications to change such an expectation. If an organization wants suppliers to design to meet the ergonomic needs of 95% of the work force, the purchasing specifications should clearly state this.

It does not make sense to continue to purchase equipment that brings hazards into the workplace which will be the causal factors for musculoskeletal injuries, or any other type of injury. It does make sense to try to have suppliers meet properly described design standards. Further, applying the hazards analysis and risk assessment methods basic to safety through design will help resolve hazardous situations in retrofitting for existing operations.

Safety Through Design in Machine Safety

The following is an excerpt, with minor modifications, from an internal communication distributed by a well-known company. It promotes the idea of designing things right the first time. This statement is an indication of how one management group has embraced the safety through design concept:

> As shown on the chart (Figure 2), the purchase or design of new machines has many facets. Usually a new machine has its need identified as a result of a business plan to introduce a new product, upgrade an existing product or increase capacity to meet existing marketplace demand. It then takes its embryonic form in the mind of the designer, estimator, engineer or machine supplier as a set of specifications to fulfill that production need, and is first put down on paper in the form of an estimated number/scope document. (In Figure 2, CER stands for capital expenditure request.)
>
> Typically, the financial costs, technical hurdles, engineering rigors and market driven schedules are the drivers in the machine fabrication processes with everything else becoming a lower priority. Little, if any, up-front time is spent on machine safety aspects. When the machine is assembled on the production floor, effectiveness and deficiencies are then noted, and there are expensive retrofits.
>
> How is the process to be improved to avoid expensive retrofits and achieve greater effectiveness? The best time to consider safety needs is when the original process

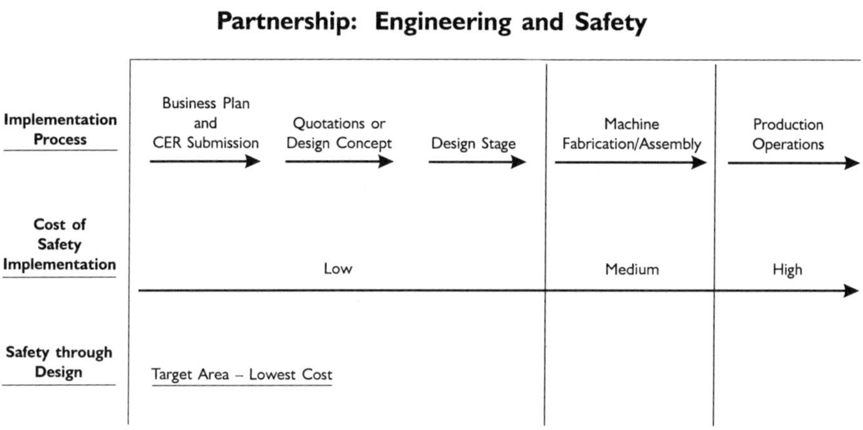

FIGURE 2 *Safety through design initiative.*

flow sheet is being determined, the preliminary machine drawings are forming, specification packages are being put together, and the bill of materials is being documented. At this stage it is just as easy to specify the correct safety interlock switch as a poor non-safety rated unit.

An example used in many Safety Through Design discussions is that it may cost as little as $1 to change a word, a specification, or a line on a designer's layout drawing, but the cost of changing these specifications could go upwards of $10,000 once the machine is in production.

What must we do? We must move from the retrofit process of machine guarding to the safety through design initiative process where we are proactive and the machines in the plant are fully compliant. We have the added advantage of doing it right the first time, and have the financial support and commitment from our management to learn and do it better.

Broader Applications of Safety-Through-Design Concepts

The examples recorded here are close to verbatim from *Safety Through Design*:

- In Chapter 13 entitled, "Application in General Industry," Paul Adams "describes a process for integrating safety into design that is being implemented and used at a major manufacturing company."
- Michael J. Douglas commences Chapter 14 with these sentences: "The first portion of this chapter covers the history of safety through design of manufacturing equipment and systems at two major automotive manufacturers, Daimler-Chrysler and General Motors. The remainder of the chapter focuses on the safety by design process within General Motors.

While the process described is specific to General Motors, it is expected that the strategy and execution would work within any structure or organization." (*Safety by design* is a General Motors term.)

- In Chapter 15 entitled, "Application in Aircraft Manufacturing," John M. Thaler states that "the integration of safety into the manufacturing design and implementation phases of new products and the modification of existing processes, when feasible, have resulted in improved injury and illness rates, productivity gains and heightened quality."
- The chemical industry has been among the leaders in applying safety-through-design concepts. In Chapter 16 entitled, "Application in the Chemical Industry," Dennis C. Hendershot provides a good discussion of the tools and techniques for safety through design in the chemical process industry.
- In Chapter 17, "Application in the Construction Industry," Craig B. Schilder claims his chapter "will provide information about two important aspects of construction design safety. The obvious one is safe design of the facility, bridge, highway, equipment or other tangible end item to ensure the user or maintenance person or disposal person is not injured. The other aspect is the adequate design to protect construction workers before the user takes possession."
- Chapter 18, "Application in the Electronic Industry," is unique in that it has two authors: Richard L. Parker, from Intel, a very large consumer of manufacturing, assembly, test and office equipment; and Mollie A. Foster from Novellus Systems, a semiconductor equipment supplier. This chapter provides a good review of how a buyer of equipment and a supplier worked together, using safety through design concepts, to keep the risks inherent in the equipment being supplied at an acceptable level.

These developments in the industries cited are indicative of the progress being made in adopting safety through design concepts. Adoption of these concepts is not limited to the industries mentioned here. Other manufacturers are using similar processes. Safety practitioners, take note.

CONCLUSION

That there is a safety through design movement in progress is a certainty. Over time, it will affect the content of the work of more safety practitioners. Those who are astute will perceive the advantage of adding the principles and methods of safety through design to their knowledge base. Also, since the data available clearly suggest that designing for safety often has a favorable impact on productivity, operating costs, and quality, the safety practitioner who applies safety through design methods will more likely be perceived as contributing to operational benefits, beyond safety.

REFERENCES

Accident Prevention Manual for Business and Industry, Administration and Programs, 11th ed. Itasca, IL: National Safety Council, 1997.

ANSI/ASQ Q9000 Series (Quality). Milwaukee, WI: American Society for Quality, revised 1994.

Christensen, Wayne C. and Fred A. Manuele, Eds. *Safety Through Design*. Itasca, IL: National Safety Council, 1999.

Christensen, Wayne C. "The Challenge: Safety through Design." *Catia Solutions Magazine*, Jan./Feb. 2000.

"Chrysler's Ergonomics Program Enters the Fast Lane." *Material Handling Engineering*, April 1996.

Control of Hazardous Energy Source (Lockout/Tagout). OSHA, Federal Register, Sept. 1, 1989, pp. 36648–36651.

Environmental, Health and Safety Performance Report. Chandler, AZ: Intel Corporation, April 2000.

Engstrom, Peter M. "Designing Machinery to Meet Global Requirements." Rept. No. 981320, Detroit, MI: Society of Automotive Engineers, 1998.

Grund, Edward V. *Lockout/Tagout—The Process of Controlling Hazardous Energy*. Itasca, IL: National Safety Council, 1995.

"Guidelines for Controlling Hazardous Energy During Maintenance and Servicing." Publ. No. 83–125, Morgantown, WV: U.S. National Institute for Occupational Safety and Health, 1983.

The Institute for Safety Through Design. Rept. to Wayne Christensen, Project Manager, Crystal Lake, IL, 2001.

ISO 14,000 Series (Environmental). Geneva, Switzerland: International Organization for Standardization, 1996.

"Lost-worktime Injuries and Illnesses: Characteristics and Resulting Time away from Work, 1997." News Release USDL 99-102, Washington, DC: U.S. Dept. of Labor, 1999.

Main, Bruce. "What Do Design Engineers Really Know About Safety?" Unpublished m.s. thesis, Ann Arbor, MI.

Manuele, Fred A. *On the Practice of Safety*. New York: John Wiley & Sons, 1997.

National Electrical Safety Code, ANSI C-2-1990. New York: American National Standards Institute.

Professional Safety. Des Plaines, IL: The American Society of Safety Engineers, June 2000.

Quality System Requirements, QS 9000, 3rd ed. Southfield, MI: Automobile Industry Action Group, 1998.

SACHE, Description of Safety and Chemical Engineering Education. Center for Chemical Process Safety of the American Institute of Chemical Engineers, New York.

Shigley, Joseph E. and Charles R. Mischke, Eds. *Standard Handbook of Machine Design*, 2nd ed. New York: McGraw-Hill, 1996.

"Sound Advise for a Quieter Workplace." *Safety + Health Magazine*, Feb. 1996.

Appendix

Mathematical Evaluations for Controlling Hazards

William T. Fine

Abstract: To facilitate expeditious control of hazards for accident prevention purposes, two great needs have been recognized. These are for:

(1) a method to determine the relative seriousness of all hazards for guidance in assigning priorities for preventive effort; and
(2) a method to give a definite determination as to whether the estimated cost of the contemplated corrective action to eliminate a hazard is justified.

To supply these needs, a formula has been devised which weighs the controlling factors and "calculates the risk" of a hazardous situation, giving a numerical evaluation to the urgency for remedial attention to the hazard. Calculated Risk Scores are then used to establish priorities for corrective effort.

An additional formula weighs the estimated cost and effectiveness of any contemplated corrective action against the Risk Score and gives a determination as to whether the cost is justified.

MATHEMATICAL EVALUATIONS FOR CONTROLLING HAZARDS

The methods described in this report were conceived and developed by the author in recognition of needs for a method of determining relative urgencies for attention to hazards, and for a simple system to give guidance as to whether the estimated cost of an engineering project to eliminate a hazard is justified.

Naval Ordnance Laboratory, White Oak, Silver Springs, MD, 1971.

These needs are satisfied by the use of formulae which weigh the varying degrees of the controlling factors.

In a few months of actual use at the Naval Ordnance Laboratory, these formulae have furnished solid foundation for safety recommendations for engineering action; they have saved many thousands of dollars by canceling costly projects which the risks did not justify; and they have given better direction to the entire safety program by indicating the relative potential seriousness of all hazards.

Since weights or values assigned to the various factors of the formulae are empirical, extended experience may indicate advisability of raising or lowering some of the criteria. However, since results are primarily for comparative purposes, relative evaluations will be valid within any organization as long as standards of judgment are consistent.

These methods are promulgated for information and for any use desired by addressees. It is believed they have universal application and should provide advancement in the state of the art of the safety engineering profession.

George B. Ball
Captain, USN

Robert Ennis
By direction

Contents

CHAPTER 1: INTRODUCTION

GENERAL. The purpose of this chapter is to illustrate the need for quantitative evaluations to aid in the control of hazards and to explain the general plan of this report.

A problem frequently facing the head of any (field type) safety organization is to determine just how serious each known hazard is, and to decide to what extent he should concentrate his resources and strive to get each situation corrected. Normal safety routines such as inspections and investigations usually produce varying lists of hazards which cannot all be corrected at once. Decisions must be made as to which ones are the most urgent. On costly projects, management often asks whether the risk due to the hazard justifies the cost of the work required to eliminate it. Since budgets are limited, there is necessity to assign priorities for costly projects to eliminate hazards.

The question of whether a costly engineering project is justified is usually answered by a general opinion which may be little better than guesswork. Unfortunately in many cases, the decision to undertake any costly correction of a hazard depends to a great extent on the salesmanship of safety personnel. As a result, due to insufficient information, the cost of correcting a very serious hazard may be considered prohibitive by management, and the project postponed; or due to excellent selling jobs by Safety, highly expensive engineering or construction jobs may be approved when the risks involved really do not justify them.

In Chapter 2 of this report, a formula is presented to "calculate the risk" due to a hazard, or to quantitatively evaluate the potential severity of a hazardous situation. Use of this formula will provide a logical system for safety and management to determine priorities for attention to hazardous situations, and guidance for safety personnel in determining the areas where their efforts should be concentrated.

In Chapter 3 of this report, a formula is presented for determining whether or not the cost of eliminating a hazard is justified. Use of the formula will provide a solid foundation upon *which safety* personnel may base their recommendations for engineering-type corrective action. It will assure that projects which are not justified will not be recommended.

This report deals with justification of costs to eliminate hazards. This does not imply in anyway that a cost, no matter how great, is not worthwhile if it will prevent an accident and save a human life. However we must also consider accident prevention with reason and judgment. Budgets are not unlimited. Therefore the maximum possible benefit for safety must be derived from any expenditure for safety. When an analysis results in a decision that the cost of certain measures to eliminate a hazard "is not justified," we do not say or suggest that the hazard 1s not serious and may a ignored.

We do say that, based on evaluation of the controlling factors, the return on the investment, or in other words, the amount of accident prevention benefit, is below the standards we have established. The amount of money involved will no doubt provide greater safety benefit if used to alleviate other higher-risk hazards which this system will identify. As for the hazard in question, less costly preventive measures should be sought.

DEFINITIONS. For the purpose of this presentation, three factors are defined as follows.

a. *HAZARD*: Any unsafe condition or potential source of an accident. Examples are: an unguarded hole in the ground; defective brakes on a vehicle; a deteriorated wood ladder; a slippery road.
b. *HAZARD-EVENT*: An undesirable occurrence; the combination of a hazard with some activity or person which could start a sequence of events to end in an accident. Examples of hazard-events are: a person walking through a field which contains a hazard such as an unguarded well opening; a person not wearing eye protection while in an eye hazardous area; a person driving a vehicle that has defective brakes; a man climbing up a defective ladder; a vehicle being driven on a slippery road.
c. *ACCIDENT SEQUENCE*: This chain of events or occurrences which take place starting with a "hazard-event" and ending with the consequences of an accident.
d. Additional definitions will be provided in later pages as needed.

CHAPTER 2: FORMULA FOR EVALUATING THE SERIOUSNESS OF A RISK DUE TO A HAZARD

GENERAL. The purpose of this chapter is to present a complete explanation of the method for quantitatively evaluating the seriousness of hazards, and some of the benefits that may be derived from such analyses.

The expression "a calculated risk" is often used as a catchall for any case when work is to be done without proper safety measures being taken. But usually such work is done without any actual calculation. By means of this formula, the risk is calculated. The seriousness of the risk due to a hazard is evaluated by considering the potential consequences of an accident, the exposure or frequency of occurrence of the hazard-event that could lead to the accident, and the probability that the hazard-event will result in the accident and consequences.

THE FORMULA is as follows:

$$\text{Risk Score} = \text{Consequences} \times \text{Exposure} \times \text{Probability}$$

$$\text{Abbreviated: } R = C \times E \times P$$

Definitions of the elements of the formula and numerical ratings for the varying degrees of the elements are given below.

a. *CONSEQUENCES C*: The most probable results of a potential accident, including injuries and property damage. This is based on an appraisal of the entire situation surrounding the hazard, and accident experience. Classifications and ratings are:

DESCRIPTION	RATING
(1) Catastrophe: numerous fatalities; extensive damage (over $1,000,000); major disruption of activities of national significance....................................	100
(2) Multiple fatalities; damage $500,000 to $1,000,000	50
(3) Fatality, damage $100,000 to $500,000................	25
(4) Extremely serious injury (amputation, permanent disability); damage $1,000 to $100,000	15
(5) Disabling injuries; damage up to $1,000................	5
(6) Minor cuts, bruises, bumps; minor damage	1

b. *EXPOSURE E*: *Frequency of occurrence of the hazard-event* (the undesired event which could start the accident-sequence). Classifications are below. Selection is based on observation, experience and knowledge of the activity concerned.

DESCRIPTION	RATING
The hazard-event occurs:	
(1) Continuously, (or many times daily)....................	10
(2) Frequently (approximately once daily)	6
(3) Occasionally (from once per week to once per month...	3
(4) Unusually (from once per month to once per year)	2
(5) Rarely (it has been known to occur)....................	1
(6) Very rarely (not known to have occurred but considered remotely possible)..........................	0.5

c. *PROBABILITY P*: This is the likelihood that, once the hazard-event occurs, the complete accident-sequence of events will follow with the necessary timing and coincidence to result in the accident and consequences. This is determined by careful consideration of each step in the accident sequence all the way to the consequences, and based upon experience and knowledge of the activity, plus personal observations. Classifications and ratings are:

DESCRIPTION	RATING
The accident-sequence, including the consequences:	
(1) Is the most likely and expected result if the hazard-event takes place	10

(2) Is quite possible, would not be unusual, has an even 50/50 chance	6
(3) Would be an unusual sequence or coincidence	3
(4) Would be a remotely possible coincidence. (It has happened here)	1
(5) Extremely remote but conceivably possible. (Has never happened after many years of exposure.)	0.5
(6) Practically impossible sequence or coincidence; a "one in a million" possibility. (Has never happened in spite of exposure over many years.)	0.1

EXAMPLES. The use of this formula is demonstrated by actual examples. Six widely different types of situations have been selected to illustrate the broad applicability of the formula and method of computation. (*In this condensation of Fine's paper, three examples are given.*)

a. *EXAMPLE NO. 1*

(1) *Problem.* There is a quarter-mile stretch of two-lane road used frequently by both vehicles and pedestrians departing or entering the grounds. There is no sidewalk, so pedestrians frequently walk in the road, especially when the grass is wet or snow covered. There is little hazard to pedestrians when all the traffic is going in one direction only; but when vehicles are going in both directions and passing by each other, the vehicles require the entire width of the road, and pedestrians must then walk on the grass alongside the road: It is considered that an accidental fatality could occur if a pedestrian steps into the road, or remains in the road at a point where two vehicles are passing.

(2) *Steps to Use the Risk Score Formula.*

Step 1. List the accident-sequence of events that could result in the undesired consequence

1. It is a wet or snowy day, making the grass along the road wet and uninviting to walk on.
2. At quitting time, a line of vehicles, and some pedestrians are leaving the grounds, using this road.
3. One pedestrian walks on the right side of this road, and he has an attitude which makes him oblivious to the traffic. (This is the hazard-event.)
4. Although traffic is "one way" out at this time, one vehicle comes from the opposing direction causing the outgoing traffic line to move to the right edge of the road.
5. The pedestrian on the right side of the road fails to observe the vehicles, and he remains in the road.

6. The driver of one vehicle fails to notice the pedestrian and strikes him from the rear.
7. Pedestrian is killed.

Step 2. Determine values for elements of formula:

- *Consequences*: A fatality. Therefore C = 25.
- *Exposure*: The hazard-event is event 3 above, the pedestrian remaining in road and refusing to notice the line of traffic. It is considered that this type individual appears or is "created" by conditions occasionally. Therefore E = 3.
- *Probability* of all events of the accident sequence following the hazard-event is: "conceivably possible, although it has never happened in many years." Reasoning is as follows: events 4, 5, 6 and 7 are individually unlikely, so the combination of their occurring simultaneously is extremely remote.

Event 4 is unlikely because traffic is "one way" at quitting time.

Event 5 is unlikely because a number of drivers would undoubtedly sound their horns and force the pedestrian's attention.

Event 6 is unlikely because most drivers are not deliberately reckless.

Event 7, a fatality, is unlikely because vehicle speeds are not great on the road, and the most likely case would be a glancing blow and minor injury. Not even a minor injury has ever been reported here. In view of the above Probability P = 0.5.

Step 3. Substitute into formula and determine the Risk Score.

$$R = C \times E \times P = 25 \times 3 \times 0.5 = 37.5$$

(*NOTE*: The Risk Score or one case alone is meaningless. Additional hazardous situations must also be calculated for comparative purposes and a definite pattern. Additional cases are similarly calculated below.)

b. *EXAMPLE NO. 2*

(1) *Problem.* Fifty (50) compressed air hoses are in use for general cleaning purposes in a machine shop, being used without proper pressure-reduction nozzles at various pressures, some up to 90 pounds per square inch. This causes potential eye hazards, although eye protection is worn by most men. The most probable consequence of his hazard is a serious eye injury.

(2) *Using the Risk Score Formula.*

Step 1. List the sequence of events to cause an eye injury accident:

- Many machine operators use compressed air streams to blow metal chips from work.
- Most employees occasionally remove their safety glasses while still in the hazardous area. (This is the hazard-event.)
- One employee who is not wearing eye protection walks past a machine while an air hose is being used.
- A metal chip blows into the employee's eye.
- A serious-eye injury results.

Step 2.

- *Consequence*: A disabling eye injury. C = 5.
- *Exposure*: The hazard-event (an employee removing his eye protection while still in an eye hazardous area) is considered to occur many times daily. E = 10.
- *Probability*: The total accident-sequence is considered "quite possible." P = 6.

Step 3. Substitute into formula and determine the Risk Score.

$$R = C \times E \times P = 5 \times 10 \times 6 = 300$$

c. *EXAMPLE NO. 3*

(1) *Problem.* A 12,000 gallon propane storage tank is subject to two hazards. One hazard is the fact that the tank is located alongside a well-traveled road. The road slopes, and is occasionally slippery due to rain, snow or ice. It is considered possible that a vehicle (particularly a truck) could go out of control, leave the road, strike and rupture the tank, and cause a propane gas explosion and fire that could destroy several buildings, with consequences amounting to damage costing $200,000, plus a fatality.

The second hazard is the tank's location close to ultra-highly compressed air lines and equipment. A high pressure pipeline explosion could result from a malfunctioning safety valve, a human error in operating the equipment, damage to a pipeline, or from other causes. Blast or flying debris could conceivably strike the propane storage tank, rupture it and cause it to explode with the same consequences as for a run-away vehicle.

(2) *Using the Risk Score Formula.* (*NOTE*: in this case there are two hazards, so the evaluation is done it two parts, one for each of the hazards, and the total scores are added.)

Step 1. Consider—just the first hazard, that due to a vehicle. List the sequence of events that would result in an accident:

1. Many vehicles are driven down the hill alongside the storage tank.
2. The road has suddenly become slippery due to an unexpected freezing rain.
3. One truck starts to slide on the slippery road as it goes down this hill. *NOTE*: This is the "hazard-event" that starts the accident sequence.
4. The driver loses his steering control at a point when he is uphill from and approaching the tank.
5. Brakes fail to stop the vehicle from sliding.
6. Vehicle heads out of control toward the tank.
7. Vehicle strikes the tank with enough force to rupture it and permit the propane gas to leak out.
8. A spark ignites the propane.
9. Explosion and conflagration occur.
10. Building and equipment damage is $200,000, and one man is killed.

Step 2. Substitute numerical values into formula:

- *Consequences*: One fatality and damage loss of $200,000. Therefore C = 25.
- *Exposure*: The hazard-event that would start the accident sequence is—the truck starting to slide on this road. This has happened "rarely." Therefore E = 1.
- *Probability*: To decide on the likelihood that the complete accident-sequence will follow the occurrence of the hazard-event, we consider the probability of each event:
  a. Loss of steering control to occur at the precise point in the road approaching the tank is possible but would be a coincidence.
  b. Once the vehicle started to slide, if the road was ice covered, it would be expected that the brakes would fail to stop the slide.
  c. The vehicle heading toward the tank is highly unlikely. Momentum would cause the vehicle to continue straight down the road.
  d. The vehicle striking the tank with great force is extremely unlikely.

If a vehicle were sliding on an ice covered surface toward the tank, it would be easily diverted from its direction of travel by a number of obstructions between the road and the tank. When roads are slippery, travel is curtailed and drivers are cautioned to drive slowly. A slow rate of speed would be unlikely to produce enough force to damage the tank. The shape and position of tank are such that a vehicle would tend to glance off it.

In summary, because of the highly unlikely nature of most of the events, this sequence has a one-in-a-million probability. It has never happened, but it is conceivable. Therefore, $P = 0.5$.

Step 3. Substitute in the formula.

$$\text{Risk Score} = 25 \times 1 \times 0.5 = 12.5$$

The entire process is to be repeated for the second hazard, which is the location near the high pressure air lines and equipment.

Step 1. List the sequence of events.

1. Normal daily activities involve operation of equipment and pressurizing of pipelines, some of which are in the vicinity of the propane storage tank.
2. A pipeline containing air compressed to 3,000 pounds per square inch, approximately 50 feet away from the storage, tank, has become deteriorated or damaged. (This is the hazard-event.)
3. The pipeline bursts.
4. Metal debris is thrown by the blast in all directions, several pieces flying and striking the propane tank with such force that the tank is ruptured.
5. Propane starts to leak out of the tank.
6. A spark ignites the propane fumes.
7. The propane and air mixture explodes.
8. Building damage is $200,000, and one man is killed.

Step 2. Determine values and substitute in the formula.

- *Consequences*: One fatality and damage loss of $200,000. $C = 25$.
- *Exposure*: High pressure air lines have been known to have been neglected or damaged. Frequency of such occurrences is considered "unusual." Therefore $E = 2$.
- *Probability*: Now we estimate the likelihood that a damaged pipeline will explode and the explosion will occur close enough and with enough blast to throw debris and strike the propane tank with such force as to complete the accident sequence. Several bursts have occurred in the past years, but none have damaged the propane tank. Few of the pipelines are close enough to endanger the tank. After careful consideration, the accident sequence is considered "very remotely possible." $P = 0.5$.

Step 3. Substituting into the formula.

$$\text{Risk Score} = 25 \times 2 \times 0.5 = 25$$

$$\text{Totaling Risk Score} = 12.5 + 25 = 37.5$$

CHAPTER 3: FORMULA TO DETERMINE JUSTIFICATION FOR RECOMMENDED CORRECTIVE ACTION

GENERAL. The purpose of this chapter is to describe the method of determining whether the cost of corrective action to alleviate a hazard is justified. Once a hazard has been recognized, appropriate corrective action must be tentatively decided upon and its cost estimated. Now the "Justification" formula can be used to determine whether the estimated cost is justified.

THE FORMULA is as follows:

$$\text{Justification} = \frac{\text{Consequences} \times \text{Exposure} \times \text{Probability}}{\text{Cost Factor} \times \text{Degree of Correction}}$$

Elements are abbreviated:

$$\text{J} = \frac{\text{C} \times \text{E} \times \text{P}}{\text{CF} \times \text{DC}}$$

It should be noted that the elements of the numerator of this formula are the same as in the Risk Score formula described in Chapter 2. We have simply added a denominator made up of two additional elements which are as follows:

a. *COST FACTOR CF:* A measure of the estimated dollar cost of the proposed corrective action: Classifications and ratings are:

COST	RATING
(1) Over $50,000	10
(2) $25,000 to $50,000	6
(3) $10,000 to $25,000	4
(4) $1,000 to $10,000	3
(5) $100 to $1,000	2
(6) $25.00 to $100	1
(7) Under $25.000	0.5

b. *DEGREE OF CORRECTION DC:* An estimate of the degree to which the proposed corrective action will eliminate or alleviate the hazard, forestall the hazard-event, or interrupt the accident sequence. This will be an opinion based on experience and knowledge of the activity concerned. Classifications and ratings are:

DESCRIPTION	RATING
(1) Hazard positively eliminated, 100%	1
(2) Hazard reduced at least 75%, but not completely	2
(3) Hazard reduced by 50 to 75%	3

(4) Hazard reduced by 25 to 50% 4
(5) Slight effect on hazard, less than 25% 6

CRITERIA FOR JUSTIFICATION. Values are substituted into the formula to determine the numerical value for Justification. The Critical Justification Rating is 10. For any rating over 10, the expenditure will be considered justified. For a score less than 10, the cost of the contemplated corrective, action is not justified.

NOTE: The critical Justification Rating has been arbitrarily set at 10, based on experience, judgment and the current budgetary situation. After extended experience at any individual organization, based on accident experience, budgetary situations, and appraisals of the safety status, it may be found desirable to raise or lower the critical score.

EXAMPLES. The use of the Justification formula will be illustrated by the use of the same six examples discussed in Chapter 2. (Three will be shown here.)

a. *EXAMPLE NO. 1.* The hazard of pedestrians and vehicles using the same road.

To reduce this risk, the corrective action being considered is to construct a sidewalk alongside the road, at an estimated cost of $1,500.00. The "J" formula is now used to determine whether, this contemplated expenditure is justified.

(1) *Substitute Values in the "J" Formula.*

$$J = \frac{C \times E \times P}{CF \times DC}$$

(a) *C, E, and P*, for this situation were discussed as Example No. 1 in Chapter 2 of this report and determined to be 25, 3 and 0.5, respectively.

(b) *Cost Factor*: The estimated cost is $1,500. Therefore CF = 3.

(c) *Degree of Correction*: The probability of the hazard-event occurring is considered to be reduced at least 75 percent, but not 100 percent, by the construction of a sidewalk: Therefore DC = 2.

(d) *Justification Rating*:

$$J = \frac{25 \times 3 \times 0.5}{3 \times 2} = \frac{37.5}{6} = 6.25$$

(2) *Conclusion.* "J" is less than 10. Therefore the cost of construction of the sidewalk is not justified.

NOTE: This lack of sufficient justification evaluates the situation from the safety viewpoint only. Management could feel there is added justification for morale or other purposes.

(3) Additional consideration. Since the Risk Score is still a substantial 37.5, other less costly corrective measures should be sought. This includes improved administrative controls to enforce one-way traffic measures, reduce speed, and encourage pedestrians to use another exit gate. This will reduce the Risk Score by reducing both Exposure and Probability.

b. *EXAMPLE NO. 2*. The hazard due to compressed air being used in a shop without proper pressure reduction nozzles.

The proposed corrective action is installation of proper pressure reducing nozzles on the 50 air hoses, at a cost of \$2.00 each, or \$400.00. To determine justification for the expenditure:

(1) Determine values for the elements of the "J" formula:

(a) *C, E, and P*, were discussed in Example No. 2 of chapter 2 and evaluated at 5, 10 and 6 respectively.

(b) *Cost Factor*: The cost of the corrective action is \$400,000, so CF = 2.

(c) *Degree of Correction*: The corrective action will reduce the hazard by at least 50 percent, so DC = 3.

(d) Substituting in the formula:

$$J = \frac{5 \times 10 \times 6}{2 \times 3} = \frac{300}{6} = 50$$

(2) *Conclusion*. "J" is well above 10. The cost of installing pressure reduction nozzles is strongly justified.

c. *EXAMPLE NO. 3*. The hazardous location of the 12,000 gallon propane storage tank.

The proposed corrective action is to relocate the tank to a place where it will be less likely to be damaged by any external source, at an estimated cost of \$16,000.

(1) Determine values for elements of the formula:

(a) *C, E, and P*, were determined in Example No. 3 of Chapter 2, to be 25, 1 and 1.5 (the two hazards combined).

(b) *Cost Factor*: Cost of relocation is \$16,000. CF = 4.

(c) *Degree of Correction*: In the very best location available, there still remains a remote possibility of damage to the tank, so DC = 2.

(d) *Substituting in the Formula*:

$$J = \frac{25 \times 1 \times 1.5}{4 \times 2} = \frac{37.5}{8} = 4.7$$

(2) *Conclusion.* Based on the established criteria, the cost of relocation of the tank is not justified.

(3) It is emphasized that the conclusion in this case that the proposed corrective action is not justified, does not mean that the hazard is of little or no significance. The Risk Score is still 37.5, and this remains of appreciable concern.

Since the potential consequences of an accident are quite severe, effort should be expended to reduce the risk, by reducing either the Exposure or the Probability, or devising other less costly corrective action. In this case it is considered that an additional steel plate barrier could be erected to protect the tank from the compressed air activities, and one or two strong posts in the ground could minimize danger from the road.

Thus the Probability of serious damage to the truck, and the Risk Score, would be considerably lessened at a very nominal cost.

Recommended Procedure For Using The "J" Formula. A convenient "J" Formula worksheet is furnished for convenience in undertaking a hazard analysis to determine the Justification Rating. Once a hazard has been recognized, the following procedure is recommended:

a. State the problem briefly.
b. Decide on the most likely consequences of an accident due to the hazard.
c. Review all factors carefully, on the scene. List the actual step-by-step sequence of events that is most likely to result in the consequences chosen. You must be specific.
d. Decide on the most appropriate corrective action and obtain or make a rough estimate of its cost.
e. Consider carefully the effect of the proposed corrective action on the hazard, and estimate roughly the degree to which the dangerous situation will be alleviated.
f. If alternative corrective measures are possible, repeat steps (d) and (e) for them.
g. Select the hazard-event: the first undesirable occurrence that could start the accident sequence.
h. Consider the existing situation carefully to determine the frequency of the

occurrence of the hazard-event: by on the scene observation, and then decide on the Exposure Rating. If in doubt between two ratings, interpolate.

i. For the Probability Rating, consider the likelihood of the occurrence of each event of the accident sequence, including the resulting injury and/or damage, and form an opinion based on the descriptive words. For example, if two unusual coincidences are required, this could be considered "remotely possible"; two "remotely possible" occurrences could be "conceivably possible" etc. If in doubt between two ratings, interpolate.

   Endeavor to be consistent. Consider the occurrence of only the same consequences which were decided on in step (b) above. For example, if you decided on consequences of a fatality, then in this step you may only consider the probability of a fatality. If you also wish to consider lesser injuries, a separate and additional computation must be made, since both the Consequences and Probability evaluations would be different. Scores should be added.

j. You have now obtained ratings for all the elements of the "J" formula. Substitute in the formula and compute the Justification Score.

k. If alternative corrective measures are being considered to alleviate the hazard, compute their Justification Scores also.

l. If there are alternative corrective measures which have acceptable Justification Scores, the most desirable from the Safety standpoint is the one which would make the greatest reduction in the Risk Score. Therefore, for each alternative, assume that the corrective measures are in effect and recompute the Risk Score. Of course this selection may also be affected by external (non-safety) considerations such as the size of investment required, the relative effects on morale, esthetics, efficiency, convenience, ease of implementation, etc.

Exception To Reliance On The "J" Formula. A highly hazardous situation may exist for which no corrective action that can be devised will give an acceptable Justification Score. Obviously in such a case, whatever corrective action is necessary to reduce the Risk Score should be taken, regardless of the Justification Score.

"J" Formula Worksheet

Problem:

Sequence of events or factors necessary for accident:

1.
2.
3.
4.
5.
6.
7.
8.

Formula Factors: *Rating*

C Consequences: ______________________ ________

E Exposure: ______________________ ________

P Probability: ______________________ ________

CF Cost Factor: ______________________ ________

DC Degree of Correction: ______________________ ________

J Justification: $J = \frac{C \times E \times P}{CF \times DC} = \frac{X\ X}{X} =$

The estimated cost of corrective action is/is not justified.

Index